Modeling in Communication Behaviour of Farmers

The Authors

Dr. Banarsi Lal is presently working as an Assistant Professor/Subject Matter Specialist (SMS) (Agricultural Extension) in Krishi Vigyan Kendra (KVK), Reasi of Sher-e-Kashmir University of Agricultural Sciences and Technology-Jammu, India. After graduating from SKN College of Agriculture, Sangaria (Rajasthan), he did his Master in Extension Education from RAU, Bikaner and Doctorate from Banaras Hindu University (BHU).He has done extensive work on "Development of communication model on farm education programmes of television in Kathua district of Jammu and Kashmir" from 2002-2007. He has won gold medal in M.Sc. and topped in qualifying exam for Ph.D. programme conducted by Institute of Agricultural Sciences, BHU,Varanasi. Dr.Banarsi has completed his M.Sc. through ICAR exam. From the last seven years he has been working in KVK, Reasi with full dedication for the farmers' community. So far he has published more than 12 high value research papers, more than 80 articles, more than 20 technical bulletins and four success stories.

Dr. Shahid Ahamad completed his B.Sc. (Ag.) from N.D.U.A.T. Kumarganj, Faizabad in 1994; M.Sc. (Ag.) Plant Pathology from C.S. Azad Unversity of Agri. and Technology, Kanpur, U.P. in 1996 and Ph.D. (Botany) from C.S.J. M. Univ. Kanpur, U.P. in 2003. He worked as a Research Associate with Prof. A.K.Sarbhoy for three years (1997-2000) in the Division of Plant Pathology, I.A.R.I., Pusa, New Delhi. He was appointed as Jr. Scientist (Plant Pathology) in 2001 in the SKUAST-Jammu at Regional Agril. Research Station, Rajouri, for seven years and then appointed as Programme Coordinator Krishi Vigyan Kendra, SKUAST-J., Rajouri-185131 (J&K) in 2008. During short tenure of twelve years, he has published 100 publications in his credit. He has published 4 books, 30 research papers in the journals of national and international repute, 20 popular articles in different magazines, 25 book chapters, 01 review/conceptual articles, 13 bulletins and 18 research abstracts and 10 Reports. He has been conferred with SPPS Fellow-2005 by the Society of Plant Protection Sciences, Division of Nematology, IARI, New Delhi. Recently he has been awarded Young Achiever Award-2011 from SADHNA, H.P. and Distinguished Service Award-2012 from Bioved Research Society, Allahabad, U.P. He has been also choosen as Councillor (2012-15) of Society of Plant Protection Sciences, New Delhi.

Dr. Dipak De is presently working as Professor and Head in Department of Extension Education, Institute of Agricultural Sciences, Banaras Hindu University, Varanasi, U.P., Dr. De did his M.Sc. (Agri.) from IARI, pre- Ph.D. from IIT, Delhi and Ph.D. from Indian Agricultural Research Institute, Pusa, New Delhi. Earlier worked as an Assistant Professor in M.L.Sukhadia University, Udaipur and Associate Professor in Rajasthan Agricultural University, Bikaner. Prof.De was the executive member of International Council of International Association for Media and Communication Research, London School of Economics, London and is President of Global Communication Research Association, Sydney, Australia. Prof. De is having more than 60 research papers, authored six books and number of book chapters.

Modeling in Communication Behaviour of Farmers

Dr. Banarsi Lal

Asstt. Professor/Subject Matter Specialist (Agricultural Extension)

Dr. Shahid Ahamad

Sr. Scientist-cum-Head

Krishi Vigyan Kendra, Reasi,
Sher-e-Kashmir University of Agricultural Sciences and Technology of Jammu
J&K

Dr. Dipak De

Professor and Head, Department of Extension Education,
Institute of Agricultural Sciences,
B.H.U., Varanasi, U.P.

2017

Scholars World

A Division of

Astral International Pvt. Ltd.

New Delhi – 110 002

Cataloging in Publication Data--DK
Courtesy: D.K. Agencies (P) Ltd. <docinfo@dkagencies.com>

Lal, Banarsi, author.
Modeling in communication behaviour of farmers / Dr. Banarsi Lal, Dr. Shahid Ahamad, Dr. Dipak De.
pages cm
Study conducted in Kathua District of Jammu and Kashmir, India.
Includes bibliographical references and index.

ISBN 978-93-86071-22-4 (International Edition)

1. Communication in agriculture--India--Kathua (District) 2. Agricultural extension work--India--Kathua (District) 3. Agricultural education--Technological innovations--India--Kathua (District) 4. Television in education--India--Kathua (District) 5. Farmers--India--Kathua (District)--Attitudes. I. Ahamad, Shahid, 1972- author. II. De, Dipak, author. III. Title.

S494.5.C6L35 2016 DDC 630.14 23

Published by : **Scholars World**
A Division of
Astral International Pvt. Ltd.
– ISO 9001:2015 Certified Company –
4736/23, Ansari Road, Darya Ganj
New Delhi-110 002
Ph. 011-4354 9197, 2327 8134
E-mail: info@astralint.com
Website: www.astralint.com

Dedicated to

The Farmers

Acknowledgement

Words are not in lexicon to express my esteem, intense and profound sense of gratitude to a modest, generous, sympathetic and affable personality Prof. Dipak De, my supervisor and chairman of the advisory committee, Professor, Department of Extension Education, Banaras Hindu University, Varanasi, U.P., India, for his inspiring guidance, constant encouragement, constructive suggestions, keen and sustained interest shown during the entire course of investigation and preparation to this research report successfully.

I am thankful to the members of my advisory committee Prof. K.N. Pandey, Department of Extension Education, Prof. R.S. Dixit, Department of Agricultural Economics and Prof. S.P. Singh, Department of Horticulture.

I express my sincere thanks to my co-supervisor Prof. G.C. Mishra, Reader, Department of Agricultural Statistic, Dr. Chandra Sen, Prof., Department of Agricultural Economics, Dr. H.P. Singh, senior lecturer, Department of Agricultural Economics, Dr. B. Jirli and Dr. K. Gadai, lecturer in Department of Extension Education, for their valuable guidance, incessant interest and constructive suggestions during the course of investigation.

I enjoy proud privilege to express my sense of indebtedness with alacrity to my cronies Satya Prakash, Sanjay Kumar Mandal, Atul Sharma, Harbhajan Singh, P.K. Thakur, Sudhir Singh, Jaideep Singh, P.K. Tiwari, Bhanu Mishra, R.P. Sahu, Kundan and others whose cooperation and help in various ways brought this task to completion.

I stoop my head before my parents, whose sincere wishes and encouragement during the course of study could make Herculean task a success.

My eyes become wet with tears of pleasure and I have not sufficient words to express my utmost gratitude to all **my elder brothers and bhabhis** (sisters-in-law) and all other elder members' great affection, constant inspiration and great moral support during my study.

Words fail to describe my heartiest love and affection of the tiny tots my **nephews** and **nieces** whose charming and innocent faces stimulated my inner urge to complete my investigation.

I pay my attribute to those who remain hidden but have always sprinkled wishes of success.

I proffer my sincere thanks from the core of my heart to Mr. D. Banerjee for his esteemed co-operation and sincere attention during the typing of this manuscript.

I express my sincere thanks to farmers of the study area for their kind co-operation in providing the desirable information.

Finally, I express my sincere indebtedness to "**THE ALMIGHTY**" who did all this through me.

Banarsi Lal

Preface

Effective communication is must to bring the desired changes among the farmers' behaviour. In order to solve the problems of the farmers there is a need to properly understand the communication process. The extensionists should have the knowledge and ability to select and use the effective extension teaching methods. Research in agricultural communication field is expanding rapidly. The book entitled as "Modeling in communication behaviour of the farmers" provides a comprehensive reading material for researchers, students, planners and extension field functionaries'. It will also assist them in providing knowledge of the mathematical communication models and enable them to change the communication behaviour of the farmers. This book can also help the research scholars to further enrich the discipline of extension education at national and international level. The book is divided in five chapters each dealing with the separate aspects. The thrust of the book is on mathematical communication model building. Several new concepts and innovations have been incorporated in the book. The authors possess a vast research and field experience in the field of extension education. The authors have endeavored to build a new mathematical communication model.

I am grateful to my Ph.D. supervisor Prof. Dipak De, Late Prof. D.K. Sujan, Dr. Shahid Ahamad, Sr. Scientist-cum-Head, K.V.K., Reasi, Dr. R.P. Sahu, Dr. J.S. Manhas for their moral support. They always encouraged for their moral support.

Valuable suggestions from the readers are always welcome for further improvement of the book.

Dr. Banarsi Lal

Contents

Chapter I

Introduction

Agriculture is the main stream of Indian economy. It directly regulates the growth of economy. The main occupation of rural people is agriculture. As the scope for bringing more area under cultivation is limited, the only possible way to increase the yield is through adoption of new and improved agricultural practices and techniques, so as to meet out the long term food grain requirement of the country.

Agriculture is the mainstay of Jammu and Kashmir's economy. About 80 per cent of the population depends on agriculture. The total geographical area of this state is 222, 236 sq. km and the population is 1,0069,917 (Census 2001).

The present age has been rightly termed as an 'information age'. Information plays an immense value in our society. Information has become an integral part of our daily life. Now people want adequate and authentic information as early as possible. The mass media namely newspaper, radio and television are catering to this important need of people. For the rapid and overall development of a country it is must that the citizens of that country are well versed with the happenings around them.

Development information and technologies generated for the farmers are of no use unless these reach to the ultimate users. It has been estimated that about 30 per cent of the technologies are being received and used by the farmers (ICAR, 1988). It is further added that the technologies generated today reach to the entire ultimate users in about 20 years.

Communication is the core activity of human association in general and progress as well as development in particular. No human life can exist in isolation. A man can survive only in society and the survival in society is possible with communication. Therefore, communication is identified as the oldest continued activity of human being since birth and goes on and on till death. More precisely, communication is the basic need of human beings and web of society which makes the survival, growth, progress and development of man possible and holds the society intact and progressive. To sum-up, communication is a vital part of personal life in the society. It is equally important in business, education, civilization, administration and other situations where people encounter with each other to satisfy their needs and wishes.

Communication maintains and animates the life. It is also a motor and expression of social activity and civilization. It leads people from instinct to inspiration, through process and system of enquiry, command and control. It creates a common pool of ideas, strengthens the feeling of togetherness through the exchange of messages and translates through into action. As the world has advanced, the task of communication has become more complex. However, unless some basic structural changes are introduced, the potential benefits of technological and communication development will hardly be put at disposal of the majority of mankind.

The rural poverty and its related incidences may decline if one puts efforts for sustained growth in agricultural production (Mellor and Desai, 1986). The extension/communication/dissemination system and network is the key input in increasing the performance in agricultural production. Therefore, the communication is the most powerful input which brings substantial development in socio-economic status of an individual.

Present Indian extension system is under numerous pressures where the extension workers, have to cater not only vast population but also to perform administrative, election, input supply and other works. Under these circumstances, it is not practically possible to serve all the farmers, all the time for all the problems when ratio of extension worker and farmer, the sender and receiver is more than 1:1000. Therefore, the potential of mass media can be exploited to serve the rural population in this direction.

Electronisation and mechanization in communication systems have provided opportunity to access the information rapidly, accurately and repeatedly. To reach the unreach modern electronic gadgets and systems have been introduced to cope-up the requirements. The government of India has realized the need and utility of these electronic equipments for rural population. Therefore, massive programmes of cyber extension, digital interactive distance learning, online networks, computers aided multimedia, internet and free online telephones etc. have been launched for the farmers. Some of the major extension technology systems and approaches are being used presently like call centre (1551), Cyber Extension, ATIC, computer-internet connectivity etc.

The use of present extension and communication technology system is based on the initiative of the farmers-the receiver itself. This is possible only when the farmer is conversant with the knowledge of handling system, approach etc. about present communication technology system as well as the positive attitude towards the system. In view of the progressive farmers, its use is judicious as they have high level of positive communication behaviour has resulted the desired results in their agricultural profession. As far as the farmers of U.P., Bihar and like backward areas are concerned they are traditionalist-hardliners and shy in nature with poor communication behaviour. They hesitate to ask recent information. The reason of poor communication behaviour is not only because of their personal weaknesses but there are number of constraints which come in the way and restrict them to make use of extension personnel and communication channels.

Communication in agriculture is not only to inform and create awareness among the farmers but also to implement new ideas that change the mode of farming. Village extension workers (VEWs) inform the farmers about the new technologies, but they are not keeping pace with the advancement of technical know-how. Secondly, the message has to travel through many stages from its source to the ultimate users. Due to this hierarchical transfer sometimes it loses its meaning and originality.

Communication is the vital aspect to change the behaviour of the receiver. As a matter of fact, no executive can be successful without communicating effectively with his superiors or subordinates. Messages could be in the form of words, symbols, signs, letters or actions. The importance of communication has been greatly emphasized by all the management experts. Communication is like a part of an individual's life as well as organizational existence. Its importance is self-explanatory and is having common experience of all as well.

Use of television as a powerful communication medium has no doubt to captivate the agriculture educators to harness its potential for reaching far across the nation. While it provides words with pictures and sound effect like movie, T.V. has the capacity to reach the largest number of people in the shortest possible time. People learn through the eyes and ears both thus, gain greater knowledge and understanding of the subject.

Although T.V. is a new medium, it is developing very rapidly. With the development and acceptance of colour in television one can expect a greater reality in this medium for future. The boom in television industries has not only affected urban masses but the rural masses are also fascinated with this media. Now this has become one of the most important media of mass communication for rural masses. It has played a major role in transferring latest technological know-how to the rural people. In India where the rural masses are isolated in villages the communication is difficult and challenging. In this situation television is one of the important sources of mass media which plays a pivotal role in reaching large number of people in no time.

Television can bring the world to our door steps within a second. This mass medium has made dissemination of news, information and entertainment possible on a scale unprecedented in human society.

It is undoubtedly one of the most versatile audio-visual aids ever developed. Admittedly, this is still a new field. There is a much to be done before television achieves its full usefulness in teaching.

Growth of Television in India

The television in India began modestly on September 15, 1959 by a UNESCO grant to study the use of T.V. as a medium of education, rural uplift and community development. In 1959 an experimental television programme was started to train personnel and particularly to discover what television would achieve in community development and formal education. Philips (India) demonstrated its use at an exhibition in New Delhi. The range of the transmitter was 40 kilometres and the audience comprised members of 180 tele-clubs which were provided free sets by UNESCO. The year 1961 witnessed educational television programmes on science for teachers. In the year 1965 entertainment programmes were introduced under pressure from manufacturers and the public. In the year 1967 Indian T.V. went into rural programmes and 'Krishi Darshan' programme for farmers in 80 villages tele-clubs in Delhi and Haryana were started. The year 1975-76 beamed educational programmes to villages through SITE. Commercial telecast for the first time was introduced in 1976. In 1977 terrestrial transmitters were put up at selected centres to extend television coverage. On August, 15, 1982, the national programme was inaugurated. In 1983 INSAT-IA India's first communication satellite was placed in geostationary orbit but failed in its operation. In 1983 INSAT-IB was successfully launched in orbit by the American Shuttle Challenger.

The transfer of science to rural people in India and gradual inoculation of scientific attitude in their everyday life, need to demonstrate in the language which will be understood and appreciated by the rural people. Television as an audio-visual medium of communication offers immense potential for penetrating technological information to remote corners of the country through the nationwide T.V. network.

Television is also considered very strong as the first stage of awareness. Apart from that, it speeds up entire process of adoption. It is considered as a credible source of information and is taken as authentic, trustworthy and prestigious medium of communication.

Deutsch (1952) said that models perform an organizing function in explaining the relationship of one part to another and in giving us an idea of the whole system. He also said that a model perform an explanatory function by pointing out in a somewhat simplified fashion how a system operates. The emphasis here is clearly on the nature of the process. Models also perform a predictive function as they enable us to predict the outcome of action and events. At the same time, several authors, *viz.*, Broadbeck (1959), Bross (1965), state that models have a somewhat ambiguous standing in the philosophy of science. Others view models as a broad concept from visualized ignorance to totally well-formulated scientific theories.

In the social sciences we can broadly classify models into two groups structured and process-oriented. On the other hand, a scientific theory may be viewed as a set of statements, including some law-like generalizations. In view of this, the present-day models of communication can be viewed as a first step towards communication theory.

As Dissanayake (1984) says, models can be best used for explanatory purposes only when we are alive to their limitations, as they are by no means a simulacrum of reality, but only a simplified image of it.

Statement of the Problem

It is beyond discussions and justifications that the communication behaviour is the basic activity of an individual through which the information is converted into its action for desired results. Katz and Lazarsfield (1955) considered communication behaviour as the respondents listening and reading habits. Fearing (1953) described it in the context of cognitive perceptual process which was conceived as dynamically related to 'need value' system. Rogers (1966) defined communication behaviour as the degree to which an individual is willing to seek information, use and advice. Newcomb *et al.* (1965) viewed communication behaviour manifested in sensitivity to information about the properties of the referent, the actualization of information such that the sender and receiver have more nearly equal information, leading to change in attitude structure and in motivating satisfaction of goals.

The communication being a social activity, communication behaviour is affected by number of social personal, economic, administrative and other variables. Therefore, the communication behaviour differs from individual to individual.

Different groups of villages are likely to respond to the same programme in different ways what is more; even a programme geared to the requirements of a specific group of people may fail to get them involved because of rural realities. It was therefore, felt necessary to study the communication behaviour of the farmers towards farm education programmes in Kathua district of Jammu and Kashmir.

To make the farming community enlightened and better skilled in the use of improved management practices, fast communication devices are required to break through the message effectively. Television possesses such a potential for conveying information effectively and promptly.

Under the back-drop of above explanation it becomes imperative to know whether the farmers would like to receive the farm massages through T.V. What is their behaviour towards agricultural programmes? What do they feel about the timing, duration, content and language of the programmes? Answers to such questions have been sought through this study in order to bring further improvements in telecasting the farm educational programmes. The investigator felt important to find out the expectation device and powerful medium of communication.

The important probing of the study will provide guidelines to the administrators and producers in further shaping of T.V. programmes, with special reference to agriculture. So realizing the importance and magnitude of the problem, the present study entitled "Communication Model on Farm Education Programmes of Television in Kathua district of Jammu and Kashmir" has been undertaken with the following specific objectives.

Objectives of the Study:

The study includes the following specific objectives:

1. To study the communication behaviour of the farmers towards farm education programmes of television.
2. To find out the preference of television in comparison to other agricultural information sources.
3. To examine the association between selected independent variables namely socio-economic status, education, size of land holding, annual income,

interpersonal interaction, innovation proneness, value orientation and achievement motivation with the communication behaviour of the farmers towards farm education programmes of television.

4. To identify the agricultural technology for farm education programmes of television using entertainment farm education format and ranking of programmes if available from television centres and respondents.
5. To develop an appropriate model for improving the communication behaviour of farmers.

Scope and Importance of the Study

Recent developments on the field of agriculture have brought numerous technologies/knowledge. The modern agricultural technologies yet to effectively serve at least a billion farmers throughout the world. With the advancement in agriculture, the communication technology is also fast changing. The electronisation and mechanization in communication systems have brought significant changes in the pattern and style of communication to the Indian farmers like cyber extension system, computer networking to make available agricultural technologies through ATICs, call centres-free telephone connectivity with No. 1551 etc.

The information technology revolution has provided huge opportunities to make easy access to information with interactive distance learning. The mechanism of computer and T.V. aided information technology help to reach the unreach. The extensive acceptability and use of high-end information and communication connectivity among reach and unreach had been experienced through the pilot projects like "Warna Wired Villages" in Kolhapur, "Info-villages" in Pondicherry, wireless in local communication technology for sugarcane growers in Cuddalore (T.N.) and like.

The research is the continuous process with changing scenario of developmental process. Numerous researches had been conducted in the field of dissemination of information, communication and extension systems on fundamental, applied and other issues of developmental process. Introduction of latest interactive communication technology among the rural population has opened a new vista of researchers to find out viability, acquaintance, accessibility, satisfaction, constraints and many more issues of the launched electronic communication technology and systems.

The present study was centered to investigate the communication behaviour of farmers towards farm education programmes of television with its three dimensions:

1. State of information input behaviour of the farmers,
2. State of information processing behaviour of the farmers and
3. State of information output behaviour of the farmers.

Communication behaviour of farmers is composite activity of farmers as a result of extension and media exposures. There are considerable researches which had been conducted on extension approaches, media, launched and like. The researches to find out the communication behaviour were conducted by assessing interaction pattern and use of extension methods and mass media (including T.V.) used by the individuals.

The present study was conducted in Kathua district of Jammu and Kashmir where the farmers receive and watch the farm education programmes. The study attempts to investigate and further evaluate masses through the telecasted messages. The present study was planned to evaluate how far the rural masses have gained knowledge about the agricultural messages through T.V. and changed their behaviour towards their adoption. The present study will also help to disseminate the agricultural information in more effective way.

T.V. is one of the most sophisticated means of mass communication media. It is serving the people of Kathua district of Jammu and Kashmir by disseminating the information in areas of agriculture, national integration, health and hygienic, entertainment programmes etc.

There exists a big gap between the information available and its dissemination in our country. There is a need to find out better and faster means of communication which will bridge the gap between the researches and their applicability.

T.V. is an ideal medium to convey information to illiterate and literate in urban and rural areas on whom it would have profound impact. As an instrumental device it is being used in variety of ways such as for direct teaching for supplementing formal education, for developing psychrometer skills, for adult education and for diffusion of agricultural know-how from etc. It is expected that the rural oriented T.V. programmes can solve the problems of inaccessibility, literacy and shortage of skilled persons in India.

Limitations of the Study

The studies in social sciences deal specifically with the behaviour of individuals. The human behaviour is a complex phenomenon which even differs from individual to individual. Therefore, no single research with all personal profile of people is possible. The present study is a study in social sciences which is also planned with those important personal profiles of the respondents visible directly and indirectly evolved in the communication behaviour. Therefore, the present study is also having, the following limitations which were experienced during the conduct of present study.

1. Though all possible precautious were taken to make the study precise, objective and reliable, yet because of limited time and resources at the disposal of investigator, the study was restricted to only four blocks out of nine blocks in Kathua district.
2. The information is based on verbal reply of the respondents since they do not have any written record.
3. Due to limited time, the study was conducted only on 150 respondents of different categories.

Layout of the Study

To make the data of present investigation presentable and understandable this thesis has been put forth in six chapters. The first chapter deals with introduction, objectives of the study, importance, scope of the study and limitations of the study. In the second chapter a brief review of literature related to studies is given. The third chapter on theoretical orientation discusses the models of communication, communication behaviour and other taxonomical concepts along with the conceptual framework of the present study. Fourth chapter highlights the methodology consisting of measurement of variables, instruments used, settings and statistical techniques. Fifth chapter deals with the findings and discussion. The sixth chapter embraces the summary and conclusion of the study. This is followed by bibliography and appendices.

Chapter 2

Theoretical Orientation

After going through the past research findings related to agricultural extension programmes as presented in the preceding chapter, a basis for theoretical orientation for the present study was formulated. It is well understood that the development of a conceptual framework makes research more meaningful.

It is essential that the theoretical concepts of the study must be made clear before generalizing the new concepts of the research studies. It also helps in developing sound scientific approach of the study. With this realization a separate chapter on "Theoretical Orientation" has been included in the present thesis. This chapter has been presented under the following headlines:

1. Terms related to the dependent variable
2. Terms related to the independent variables
3. Theoretical frame work.

1. Related to the Dependent Variable

a) Communication

The word communication is derived from the Latin word 'communis' meaning to develop commonness. The purpose of communication is to establish commonness between the sender and the receiver and his response.

Lasswell (1949) defined communication as "Who says what, in which channel, to whom, with what effect".

Stevens (1950) defined "Communication is the discriminatory response of an organism to a stimulus".

Hovland *et al.* (1953) defined "Communication as the process by which an individual, the communicator transmits verbal stimuli to other individual to modify his behaviour, the communicatee".

Schramm (1954) said "Communication stems from the word 'communis' meaning common. When we communicate, we try to establish' commonness' with someone. An attempt is made to share information, or an idea or an attitude. In essence it is the act of getting a sender and receiver tuned together for a particular message or a series of messages".

Rogers and Yost (1960) said "Communication in agricultural situation is the flow of research findings from agricultural scientist to the farmers".

Leagans (1961) viewed communication as the process by which two or more people exchange ideas, facts, feelings or impression in ways that each gains a common understanding of the meaning, intent and use of messages.

Ban and Hawkins (1988) defined communication as the process of sending and receiving messages through channels which establish common meaning between a source and a receiver.

Communication is conceptualized as the flow of information from different sources to ultimate users.

b) Behaviour

Behaviour is response to stimuli.

Leagans (1961) refers to what an individual knows (Knowledge), what he can do (skill) mental and physical, what he thinks (attitude) and what he actually does (action).

Peter (1977) defined behaviour as the way in which an organism interacts, shares ideas and feelings and perceives others mode of action.

Chitambar (1981) stated that concrete form of acquired new knowledge, skills and attitude through experience that provides direction to interact in the society is termed as behaviour.

Robert and Donn (1988) revealed that social behaviour is performed by specific persons, such behaviour always occurs against a back drop of socio-cultural factors (e.g., group membership, culturally shared standards and values). But their major interest is that of understanding the factors that shape and direct the actions of individual human being in a wide range of social settings.

Therefore, we can say that the results emerged from gained knowledge, skills and attitudes by an individual to interact in his society is known as behaviour.

c) Communication Behaviour

Communication behaviour is the process of acquisition, processing and dissemination of scientific information. It is found in an individual, group, organization level or in a social system. Communication behaviour deals with a communication transaction between two individuals performing their respective roles.

Katz and Lazarsfield (1955) consider communication behaviour as the respondent's listening and reading habits.

Berlo (1960) felt that how, why, when, with whom and what consequences man behaves, should come under the purview of communication behaviour.

Rogers (1966) defined communication behaviour as the degree to which an individual is willing to seek information and advice.

Thayer (1968) suggested four levels (i) the interpersonal level (ii) the intrapersonal level (iii) the organizational level and (iv)the technological level at which communication takes place.

Jha (1974) operationalised communication behaviour as responses to communication stimuli namely source, channel and message.

Murthy and Singh (1974) operationalised the receivers communication behaviour as a composite measure of awareness, comprehension, attitude and adoption.

Communication behaviour is conceptualized as integration of three main components which are information input behaviour, information processing behaviour and an information output behaviour of an individual.

(i) Information

Information is processed data which is critical input.

Webster's dictionary defined information as knowledge communicated by others or obtained by personal study and investigation, or alternatively as knowledge of a special event, situation or the like. Hence, all information is knowledge.

Leagans (1971) defined information as a message a communicator wishes his audience to receive, understand, accept and act upon. He further stated that information might consist of statement of scientific facts, description of actions being taken, reasons of why certain action should be taken, or steps necessary in taking action. Thus, information is generalized ideas, opinion, fact, feeling piece of knowledge that proceed action.

Commings (1971) stated that prime input for modernization of agriculture, is useful technical information with an adequate scientific base.

Oliver and Peter (1989) defined information as follows:

(a) Knowledge communicated or received concerning a particular fact or circumstances, news etc. is called as information.

(b) An indication of the number of possible choices or messages (in communication theory).

(c) Any data that can be coded for processing by a computer or similar device (in computer technology).

(ii) Information Input Behaviour

Information input behaviour refers to all the activities performed by an individual for acquiring scientific and technical information from various sources for performing his role effectively.

(iii) Information Processing Behaviour

Information processing behaviour defined by Thayer (1968) conceptualized information processing as a composite of information evaluation, information storage and information transformation. He further explained that information evaluation i.e., diagnosis, analysis and synthesis or decisioning, storage of information is noting, indexing and categorizing, where as the information transformation means reproduction in suitable form, amplification or reduction of initial information. Hence, the information processing refers to all the activities performed by an individual for synthesis, evaluation, storage and transformation of scientific and technical information.

(iv) Information Output Behaviour

Information output behaviour refers to all the activities performed by an individual for dissemination of scientific information to perform his role effectively.

(v) Communication Model

Models are symbolic representations of structures, objects or operations. They are useful theoretical constructs that are frequently used in social sciences for explanatory purposes. They may be used to show the size, shape or relationship of various parts or components of an object or process. A model may also be useful in explicating the working of a system.

Deutsch (1952) says that models perform an organizing function in explaining the relationship of one part to another and in giving us an idea of the whole system.

As Dissanayake (1984) says, models can be best used for explanatory purposes only when we are alive to their limitations, as they are by no means a simulacrum of reality, but only a simplified image of it.

(vi) Mathematical Models

In the category of mathematical models, the earliest work as well as the least amount of work has been done in communication research. Although their theory has served as a landmark in communication research for many years, Shannon and Weaver's writings are difficult to understand. For example, Shannon formula for information (1948) is

$$H = -[P_1 \log P_1 + P_2 \log P_2 + - - - - P_n \text{Log } P_n]$$

$$H = -\Sigma P_i \text{ Log } P_i$$

In attempting to describe information as the reduction of uncertainty, Shannon uses the term entropy and in the formula H is the mathematical symbol for entropy, S is the symbol for "sum of", Pi is the probability of an event i occurring and log P_i is the information needed to predict the occurrence of event i.

2. Related to the Independent Variables

(i) Socio-economic Status

By social status, Naik (1952) means the social position of a person or a group in the total society, consorting of a number of higher-lower positions.

According to Dictionary of sociology, Economic status means position or standing judged by a standard of wealth or lack of wealth.

Chapin, defined socio-economic status as the position of an individual or a family occupies with reference to the prevailing average standard of cultural possessions, effective income, material possession and participation in the group activity of the community.

Farber and Osoinach (1959), formulated an index of socio-economic rank on the basis of occupation, education, income and non-white population (caste).

According to Driver (1954), socio-economic status is a term based on age, religion, income and education.

The seventh evaluation report of the programme evaluation organization of the planning commission, government of India (1960) points out socio-economic factors like caste system, backwardness of the people, small holdings and others as affecting the development work in villages.

The development in the approach to the agricultural adjustment problems in Canada have resulted in enlargement of the scope of socio-economic research to emphasis social system (i.e. family, school, church, etc.) as unit of analysis in the study of communities.

The concept of socio-economic status has also tended to include the level of living status although Reid (1947) have defined the level of living as the way the groups actually live. The terms "Socio-economic status", "Level of Living" and "Plan of Living" have often been used loosely and interchangeably.

Social status is an indication of one's position of command and it may either inhibit or enhance a person's decision making ability (Rogers, 1962).

It refers to the position a farmer occupies in comparison to others with respect to possession of land, education, type of house, material possession, farm power and social participation.

It has been operationalised as the measurement developed by Trivedi (1963). Hence, it is hypothesized that farmer's socio-economic status would be directly associated with their communication behaviour.Higher the socio-economic status, higher the communication behaviour of the farmers.

(ii) Education

According to Webster dictionary(1963) education means discipline, development and training of the mind, character and faculties, instruction and training of young, organized system of instruction as existing in a given state.

According to international encyclopedia of the social sciences, education as an institutionalized form of socialization to adult roles.

Educational groups are both formal and informal. The formal is well organized and characterized by informative action. The informal is unorganized, indirect and characterized by personal appeal.

In modern society education is more formal, systematically organized and bureaucratized. And in most modern societies, it is compulsory, there is no other recognized path by which the average child may be socialized and awarded an achieved status.

According to Beal and Sibley (1967) the individuals' ability to read and write and the amount of formal education he possesses will effect the manner in which the individual gather data and relates himself to his environment.

Therefore, it can be inferred that education helps to an individual to know and give shape and direction to their thinking process about the message. It affects the process of communication behaviour. It has been operationalised as the formal education obtained from the University by the individual respondents.

In the present study, education has been conceptualized as measured by Trivedi scale, 1963. Past researches have shown that education as a variable plays an important role in influencing the communication behaviour of the farmers. Therefore, it has been hypothesized that level of education of respondents would be associated with the communication behaviour of the farmers. Higher the education, higher the communication behaviour of the farmers.

(iii) Size of Land Holding

It refers to the total land owned by farmers. It was inclusive of the area under houses, farm yard, pastures, net cultivable land and any other type of land owned by farmers but not the leased land. It was measured in acres.

Size of land holding denotes actual area of land owned by an individual family in terms of standard acres. Studies on communication behaviour indicate that size of land holding is positively and significantly related with the communication behaviour of the farmers. Size of land holding is also an indicator of economic status in the rural area. The number of acres of land under cultivation was taken as the measure of size of land holding.

In the present study, size of land holding has been conceptualized as measured by Trivedi scale, 1963. Past researches have shown that size of land holding as a variable plays an important role in influencing the communication behaviour of the farmers. Therefore, it has been hypothesized that there is a relationship between size of land holding and communication behaviour of the farmers.

(iv) Annual Income

It refers to the total income in rupees earned by the respondents from all sources in a particular year.

A rapidly growing population is associated with high percentage of young persons in the age structure of the population. Children usually do not contribute much to production, but they do participate in consumption. The higher so-called dependency ratio- the proportion of economically non-active to economically active persons - in economy, the more the active members of the labour force have to produce to attain a given level of welfare. Several theories state that population growth, working through the dependency ratio, has a negative effect on overall economic development.

If there are many small children in a household, the consumptive expenditures of the household will tend to be greater and the proportion of household income saved will be smaller. A well-known hypothesis states that a 'high dependency ratio' will result in a low savings rate in an economy. Furthermore, many children in poor families after result in lower consumption, rather than lower savings. Finally, large number of children can provide incentives to work harder and to produce more in order to be able to feed more mouths. Nevertheless, the World Bank concludes tentatively that a large family size forms a heavy economic burden for a family and may indirectly have negative effects on its long run capacity to save.

Annual income is also an indicator of economic status in the rural areas. Thus, in the present investigation, annual income has been conceptualized and measured in terms of rupee earned by the respondents from all sources. Past researches have shown that annual income as a variable plays an important role in the use of various sources of information and influencing the communication behaviour of the farmers.

(v) Interpersonal Communication

Barnlund (1968) "The study of interpersonal communication is concerned with the investigation of relatively internal social situations in which persons in face-to-face encounters sustain a focused interaction through the reciprocal exchange of verbal communication".

Bettinghans (1968) at the simplest level, a communication situation exists whenever one person transmits a message that is received by another individual and is acted upon the individual. Thus as human communication situation includes individual 'A' a source of communication using symbols or stimuli that have shared meanings of individuals as a message to be delimited or passed along some channel to individual 'B' a receiver of communication.

Katz and Kahn (1966) generally concluded that in a well functioning system, interpersonal communication must flow both ways freely and that informal communication by passes and parallels the formal hierarchical pattern.

Von Blackenburg (1976) maintained that in most rural areas of developing countries the social disparity could be minimized through maximizing interpersonal communication.

According to Dhama and Bhatnagar (1980) in a face to face situation, communication is not mere exchange of information but something more, because in such a situation, along with the information on passes, the gestures, expression, language, the manner of expression and tone-all these combined together, create a sort of impact on both. Some kind of change occurs as a result of interaction. This change may be visible in interaction of knowledge and behaviour. In the present study, interpersonal communication has been conceptualized as measured by Bhople (1985). Hence, it is hypothesized that interpersonal interaction would be directly associated with communication behaviour of receivers. Higher the interpersonal communication, higher the communication behaviour of the farmers.

(vi) Innovation Proneness

It is the degree of an individual interest and a desire to seek change in farming techniques and to introduce such changes into their operations when practical and feasible (Maulik, 1965). In the present study, innovation proneness has been conceptualized as measured by Choudhary. Hence, it is hypothesized that innovation proneness would be directly associated with communication behaviour of receivers. Higher the innovation proneness, higher the communication behaviour of the farmers.

(vii) Value Orientation

According to Wilkening (1954) values are abstract concepts inferred from behaviour, operate to influence a selection of available means and ends of action and have either favourable or unfavourable connotations for the well being of an individual or of the group.

According to Parsons (1961) a value is a normative pattern which define desirable behaviour for a system in relation to its environment, without differentiation in terms of the function of units or of their particular situations.

English and English (1961) defined value as abstract concept, often merely implicit, that defines for an individual or for a social unit what ends or means to an end are desirable. They further emphasized that these abstract concepts of worth are usually not the result of individuals own valuing, they are social products that have been imposed on him and slowly internalized i.e., accepted and used as his own criteria of worth.

Speaking to value, Dube (1963) said that organized round the major themes of culture, value delimit the parameters of action by ascribing most desirable, desirable neutral, undesirable and most undesirable qualities to possible choices in a given situation. In other words, they are a series of explicit or implicit culturally sanctioned guides to action that set the direction and units of behaviour in specific situation within the frame work of a given culture.

According to Lionberger (1964) values may be regarded as important ratings which people attach to things, conditions and circumstances. They may also be regarded as goal objects to which people orient their thinking, actions and feelings. As such they become important organizing themes in the behaviour of individuals.

Then for the purpose of the present study value may be regarded as an abstract concept which serves as a guide to communication behaviour and for the selection of available means and ends of action.

In the present study, value orientation has been conceptualized as measured by Kittur (1976). Hence, it is hypothesized that value orientation would be directly associated with the communication behaviour of the farmers. Higher, the value-orientation, higher the communication behaviour of the farmers.

(viii) Achievement Motivation

Murray (1938) conceived achievement motivation as a desire or tendency to do things as rapidly and/or so well as possible. He specified the desire as to accomplish something different to master, manipulate, or organize physical objects, human beings or ideas, to overcome obstacles and attain a high standard to excel one's self, to rival and surpass others.

McClelland *et al.* (1953) stated that "Success in competition with some standard of excellence is need achievement". They characterized achievement motivation as 'the propensity to strive for success in any and all situations in which a standard of excellence was thought to apply'.

Edwards (1957) defined achievement motivation to do one's best, to be successful, to accomplish tasks requiring skills and efforts to be a recognized authority and like.

Atkinson (1958) conceived achievement motive as latent disposition to strive for particular goal, state or aim.

McClelland (1961) proposed a model of achievement motivation, usually referred to as A-T-D model. This model is brief, can be explained as that a raise in the level of ç-Ach (A) in a society will be associated with a rise in the rate of development (D) with a time lag (T) which will be always positive.

Rogers and Svenning (1969), defined achievement motivation as a spontaneously expressed desire to do something well for its own sake rather than to gain power or love or recognition.

In the present study, achievement motivation has been conceptualized as measured by Visweshwaram (1969). Hence, it is hypothesized that achievement motivation would be directly associated with the communication behaviour of the farmers.Higher,the achievement motivation, higher the communication behaviour of the farmers.

3. Theoretical Frame Work

The present scientific investigation into the communication behaviour of the farmers on farm education programmes of television closely associated with the modernizing, diversification and intensification of agriculture, calls for a sound system based on theoretical frame work.

There has always been a gap, between knowledge production and knowledge utilization. This is because the language of science is not the same as that of its integrators. Researchers in any given field such as agriculture and the client in that field (farmers in case of agriculture) present in two different social systems, each identified by its norms, values, goals, culture, language and communication pattern (Havelock, 1968). These norms within each system also define their separateness from each other. The inadequacy of shared characteristics between the innovation development system and the client system lead to serious communication gap between science and practice. While analysing the problem of bridging the gap, *Guba* (1968) concluded that it cannot be spanned either by the producer of the knowledge or by its utilizer himself, or even by these two acting in concert, at least in the typical situation.

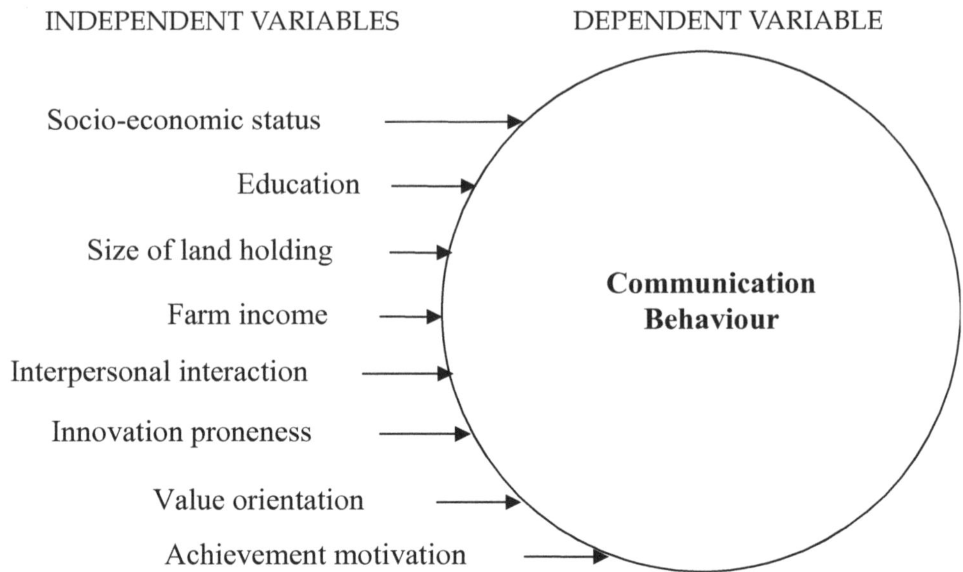

Figure 1: Theoretical Model of the Study.

It was realized long back, in almost every field that some sort of mechanisms or agencies, using special techniques be interposed to perform this bridging or limiting function. Accordingly, various kinds of limiting system or some such mechanisms have emerged in the fields like agriculture, education, public health, various industrial enterprises etc.

T.V. is an ideal medium to convey information and news to illiterate and literate, urban and rural viewers on whom it would have profound impact. As an instructional device it is being used in variety of ways i.e., for direct teaching for supplementary formal education, for developing psychometer skills, for adult education and for diffusion of agricultural know how etc. It is expected that the rural oriented T.V. programmer can solve the problems of inaccessibility, literacy and shortage of skilled persons in India.

In the tentative model, there are eight independent variables that might be associated with the communication behaviour of the farmers towards farm education programmes of T.V. As shown in model below it was presumed that these variables might affect the communication behaviour of the farmers towards farm education programmes of T.V.

Chapter 3

Research Methodology

To undertake any research programme a well defined methodology is necessary in order to fulfill its objectives. A review of earlier studies provides a good practical and theoretical background for designing the present study and facilitates in decision making in the selection of suitable methodology and appropriate tools in accordance with the objectives of the study. The methodology is described under the following sections:

1. Locale of study
2. Selection of sample
   (a) Selection of Community Development (C.D.) Blocks
   (b) Selection of Gram Panchayats
   (c) Selection of villages
   (d) Selection of respondents
3. Variables and their empirical measures
   (a) Measurement of dependent variable
   (b) Measurement of independent variables
4. Techniques of data collection
5. Formulation of Hypotheses
6. Statistical analysis of data

1. Locale of Research

The study was conducted in Kathua district of Jammu and Kashmir. Kathua district touches with the boundaries of Himachal Pradesh and Punjab. The soil ranges from sandy loam to alluvial soil and the main crops grown in the district are wheat, maize and paddy. About 80 per cent of people of state depend on agriculture. Literacy of the state is 54.46 per cent and J and K consists of 14 districts. The total geographical area of Kathua district is 2,651 sq. km. and the population is 544, 206 (census 2001). Out of 14 districts Kathua district has selected purposively because of the following reasons.

1. All the agricultural based programmes telecast by Delhi and Jalandhar (Punjab) Doordarshan Kendras are received and viewed by the farmers of this district.
2. There is a K.V.K. located in Kathua district and the farmers of this district are seeking agricultural information from the K.V.K.
3. There is a transmitting T.V. Centre in Kathua district.
4. The investigator belongs to Kathua district and so he is well versed with the local dialect, culture and background of people. This helped the researcher to collect authentic and reliable data.

2. Selection of Sample

a) Selection of C.D. Blocks : Kathua district consists of 9 blocks. Out of which 4 blocks namely Ghagwal, Hiranagar, Barnoti and Kathua were selected randomly.

b) Selection of Gram Panchayats Ghagwal, Hiranagar, Barnoti and Kathua comprise 18, 24, 30 and 22 Gram Panchayats respectively. 20 per cent of the Gram Panchayats from each block were selected randomly for the study purpose.

c) Selection of villages: A complete list of the villages from each Gram Panchayat was made and a sample of 20 per cent villages was selected randomly for the study purpose.

d) Selection of Respondents: A list of the farmers possessing T.V. from selected villages was prepared. A sample of 20 per cent respondents was selected randomly for the study purpose.

Table 1: Distribution of Selected Villages

Name of Gram Panchayat	*Total No. of Villages*	*No. of Selected Villages*	*Name of Selected Villages*
Chhankhatrian	4	1	Chhankhatrian
Ghagwal	4	1	Ghagwal
Kattal	9	2	Kattal BrahmanaRaghu Chak
Londi	8	2	Katal GujranLondi
Bobiya	8	2	BobiyaLadwal
Chadwal	3	1	Chadwal
Chak Bhagwna	8	2	Chak BhagwnaKhanak
Devo Chak	3	1	Devo Chak
Jandi	6	1	Jandi
Chandwan	3	1	Chandwan

Contd...

Table 1–*Contd...*

Name of Gram Panchayat	*Total No. of Villages*	*No. of Selected Villages*	*Name of Selected Villages*
Khanpur	10	2	KhanpurPariyal
Chak Karna	4	1	Chak Karna
Mukandpur	5	1	Chak Shama
Chak Nathal	8	2	Chak NathalMalwan
Uttri	4	1	Uttri
Gobindsar	8	2	Ch. HarisinghGobindsar
Basantpur	8	2	BarniDanna
Khukhyal	5	1	Khukhyal
Rakh Lachipur	5	1	Rakh Lachipur

3. Variables and their Empirical Measurement

The selection and measurements of the variables for the present study were based on the objectives of the study and the guidelines received from the literature and the experts involved this field. The variables selected and their empirical measurements are presented in the following table:

Table 2: Variables and their Measurement

Sl.No.	*Variables*	*Empirical Measurement*
1(a)	Dependent variable Communication behaviour	Communication behaviour index was developed for the study
(b)	Independent variables	
1)	Socio-economic status	Trivedi, 1963
2)	Education	Trivedi, 1963
3)	Size of land holding	Trivedi, 1963
4)	Annual Income	Index was developed
5)	Interpersonal interaction	Scale developed by Bhople (1985)
6)	Innovation proneness	Scale developed by Choudhary (1983)
7)	Value Orientation	Scale developed by Kittur (1976)
8)	Achievement motivation	Scale developed by Visweshwaram (1969)

Definition and Operationalization of Concepts

A lot of terminologies have been used in this study with specific meanings and concepts. According to Kerlinger (1965) operational definition is a specification of the activities of the researchers in measuring a variable or in manipulating it. He further suggested that an operational definition is a sort of manual of institutions to the researcher. Roger and Shoemaker (1971) stated that it is a scale, index, observation or the answer to a direct question. This makes the meaning of a concept more explicit.

The various concepts used in this study have been operationalized in explaining their epistemic relationship and empirical measurement.

Communication

The word communication is derived from the Latin word 'communis' meaning commonness. According to Thayer (1968) 'communication is the natural interchange of ideas by any effective means'. Leagans (1961) defined communication as a process by which two or more people exchange ideas, facts, feelings or impressions in ways that each gains a common understanding of meaning, intent and use of message. For the present study, communication viewed as flow of information from sources into the ultimate users and vice-versa.

Communication Behaviour

Katz and Lazarsfield (1955) considered communication behavior as the respondents listening and reading habits. Rogers (1968) defined communication behaviour as the degree to which an individual is willing to seek information and advice. Jha (1974) operationalized communication behaviour as responses to communication stimuli namely, source, channel and message. Communication behaviour at the individual level analysis can be analyzed in terms of three activities (1) Information input, (2) Information processing and (3) Information output.

Information Input Behaviour

It refers to all the activities performed by an individual (farmer) for the acquisition of scientific information, technical know how and innovative ideas from different sources is known as 'information input behaviour. The similar pattern of operationalization of this concept was also used by Jain (1970), Akhouri (1973), Shete (1974), Ambastha (1974) and Sanoria (1974). The amount of information input of the respondents was measured with the help of information input index developed for them. Each of the item of the information input index was assigned numerical score arbitrarily. Due to variation in the range as well as units of measurements, the scoring was made uniform for all the items. included in the information input index. Based on this assumption, the real score for each of the items of the information input index was determined by using the following formula :

$$\text{Real score of an item} = \frac{\text{Obtained score}}{\text{Maximum possible obtainable score}} \times 100$$

The real score of all the items were summed up to obtain information input index of individual.

Information Processing Behaviour

Thayer (1968) viewed that information processing deal with evaluation of the information received. It has met the analysis, synthesis and decisioning or cataloguing. Transformation is reproduction in suitable form, amplification or reduction of initial information. In this study, information processing behaviour refers to all the activities performed by an individual (farmer) for evaluation, storage of scientific and technical information that is reproduction in suitable form. Each item has numerical values. The score of each item was converted into real score of an item by applying the formula described under information input index. The real score of all these items were summed up to obtain information processing index of individual.

Information Output Behaviour

Ambasta (1974) referred it to the knowledge, attitude, use, adoption, dissemination of information to follow farmers and transmission of farm problems to extension personnel or researcher. In this study output pattern is restricted with the dissemination of information to others.

Each item had different values. The raw score of each item was converted into real score of an item by applying the formula, described under information processing index. The real score of all these items were summed up to obtain information output index of individual.

Communication Models

Models are symbolic representations of structures, objects or operations. They are useful theoretical constructs that are frequently used in social sciences showing the size, shape or relationship of various parts or components of an object or process. A model may also be useful in explicating the working of a system.

Deutsch (1952) says that models perform an organizing function in explaining the relationship of one part to another and in giving us an idea of the whole system. He also says that a model performs an explanatory function by pointing out in a somewhat simplified fashion how a system operates.

As Dissanayake (1984) says, models can be best used for explanatory purposes only when we are alive to their limitations, as they are by no means a simulacrum of reality, but only a simplified image of it.

Independent Variables

(a) Socio-economic status

To measure the socio-economic status, the scale developed by Trivedi (1963) was found suitable for present study. The scale consists of caste, occupation, education, social participation, land, farm power.

(b) Education

Education scores for different levels were given according to the socio-economic status scale of Trivedi (1963) and these were scored.

		Score
Illiterate	–	0
Can read only	–	1
Can read and write	–	2
Up to primary level	–	3

Up to middle level	–	4
High School	–	5
Graduate	–	6
Above graduate level	–	7

(c) Size of Land Holding

It was operationalized for the study as the total area owned or cultivated by the respondent family. This refers the ownership of land of a respondent, which was measured as per the scale developed by Trivedi (1963). The land has been categorized in eight groups, where no land scored one, less than one ha. scored two; 1-2.5 ha. scored three, 2.6-4.0 ha. scored four, 4.1-6.0 ha. scored five, 6.1-8.0 ha. scored six, 8.1-10.0 ha. scored seven and to more than 10 ha. scored eight.

(d) Annual Income

It is the some of net annual income of the farmers obtained from the agricultural produce. The score on farm income is assigned in following ways:

Category	*Score Assigned*
Upto Rs. 5000	1
Rs. 5001 to Rs. 10000	2
Rs. 10001 to Rs. 15000	3
Rs. 15001 and above	4

(e) Interpersonal Interaction

It refers to face-to-face interaction among individuals. It has been measured through the scale used by Bhople (1985). It has been scored as follows :

Village Level Worker (VLW)	1
Exhibition	1
Training	1
Meeting	1
Posters	4
Printed Bulletin	3
Cyclostyle	1

(f) Innovation Proneness

It is the degree of an individual interest in the desire to seek change. Innovation proneness of respondents was measured through scale used by Chaudhary (1973). It has been scored as follows:

(i) (a) I try to keep myself upto date with information on new farm practices but that does not mean that I try out all the new methods on my farm.

Most like you (2) Least like you (1)

(b) I feel restless till I try out a new farm practice, I have heard about

Most like you (3) Least like you (1)

(c) They talk of many farm practices these days, but who knows if they are better than the old ones

Most like you (1) Least like you (1)

(ii) (a) From time to time I have heard of several new farm practices and I have tried out most of them in last few years.

Most like you (3) Least like you (1)

(b) I usually wait to see what results my neighbour obtain, before I try out the new farm practice

Most like you (2) Least like you (1)

(c) Somehow I believe that the traditional ways of farming are the best

Most like you (1) Least like you (1)

(iii) (a) I am cautious about trying a new practice

Most like you (2) Least like you (1)

(b) After all, our forefathers were wise in their farming practices and I do not see any reason for changing these old methods.

Most like you (1) Least like you (1)

(c) Often new farm practices are not successful however, if they are promising I would surely like to adopt them.

Most like you (3) Least like you (1)

(g) Value Orientation

Values largely influence the individual behavioural pattern. Value is the relative importance people attribute to different objects, phenomena and circumstances. People orient their thinking, feeling and actions towards different things in life, based on the value they hold. As such, they may become important organization themes in the behaviour of individuals.

For the present study, the value orientation was measured with the help of scale developed by Kittur (1976) have been used. To know the values held by the farmers four dichotomies of values were selected. They were cosmopoliteness, localiteness, skepticism, fatalism, liberalism, conservation, high aspiration, low aspiration.

This scale consists of eight items, of which four are positive, four are negative on a three point continuum. The response categories of 3, 2, and 1 for positive and reverse for negative statements. The score obtained for each statement were summed up to get individual respondents value orientation.

(h) Achievement Motivation

It was operationalised as the desire for excellence to attain a sense of personal accomplishment. It was measured with the help of a scale developed by Visweshwaram (1969).

The instrument consisted of 6 statements and responses obtained on 3 point continuum namely 'agree, 'undecided' and disagree'. A weight-age of 3, 2, and 1 respectively were assigned to the response categories in the case of positive statements and the scoring was reversed for negative statements. The total score of the respondents on their achievement motivation was arrived by summing up the weightages of responses for each statement.

4. Tools and Techniques of Data Collection and Procedure

Structured interview schedule as per the objectives of the study were prepared in consultation with the experts in this field for necessary data collection.

Interview Schedule

The main tool used for collection of data for the present study was interview schedule. Keeping in view the specific objectives and selected variables of the study, schedule was prepared for collection of necessary data. Before administration of the interview schedule, a pre-testing of the developed schedule was done among the block level functionaries of the selected sample in order to eliminate business in the schedule.

5. Formulation of Hypotheses

According to Goode and Hatt, hypothesis is, "A proposition which can be put to test to determine validity".

Considering the importance of the factors selected to be studied with reference to the objectives of the present study mentioned in chapter I, the hypotheses framed for this study have been described in null form.

HO_1: There is no relationship between socio-economic status and communication behaviour of the farmers towards farm education programmes of television.

HO_2: There is no relationship between education and communication behaviour of the farmers towards farm education programmes of television.

HO_3: There is no relationship between land and communication behaviour of the farmers towards farm education programmes of television.

HO_4: There is no relationship between annual income and communication behaviour of the farmers towards farm education programmes of television.

HO_5: There is no relationship between interpersonal interaction and communication behaviour of the farmers towards farm education programmes of television.

HO_6: There is no relationship between innovation proneness and communication behaviour of the farmers towards farm education programmes of television.

HO_7: There is no relationship between value orientation and communication behaviour of the farmers towards farm education programmes of television.

HO_8: There is no relationship between achievement motivation and communication behaviour of the farmers towards farm education programmes of television.

6. Statistical Analysis

The collected data were quantified by giving scores to each appropriate answer as referred in earlier pages. Further, the statistical tests were applied in light of the objectives to arrive at conclusions. The following statistical tools were used in the study for precise and meaningful analysis and interpretations of the quantified data.

(a) Frequency and Percentage

(b) Arithmetic Mean ($\overline{X}$)

(c) Standard Deviation (S.D.)

(d) Correlation coefficient (r)

(e) Multiple Regression coefficients

(a) Frequency and Percentage

Frequency was used to distribute the respondents as per their frequencies over different categories of their selected independent and dependent variables of the study. It was measured as the number of responses in a particular category. Percentage was used to check the frequency.

(b) Arithmetic Mean ($\overline{X}$)

Arithmetic mean is defined as the sum of all values of observations divided by the total number of observations. Mean is a measure of central tendency of the observed phenomena. Symbolically, it is represented as:

$$\text{Arithmetic Mean } (\overline{X}) = \frac{\sum X}{N}$$

where,

$\overline{X}$ = Arithmetic Mean

ΣX = Sum of all the values of observations

N = Total number of observations in the sample

(c) Standard Deviation (S.D.)

Standard deviation is the square root of the mean squared deviation. This was used to find out the variations in the scores on selected independent variables and also used for categorization of respondents. Symbolically, it is represented as:

$$\text{S.D.}\ (\delta) = \sqrt{\frac{1}{N-1} - \left(\sum X_i^{\,2} - \frac{\left(\sum X_i\right)^2}{N}\right)}$$

where,

ΣXi^2 = Sum of squares of values of the variables.

ΣXi = Sum of values of the variables

N = Number of respondents.

Mean and standard deviations observed in reference to the measurement of the important variables categorize the respondents into three different standard categories of high, medium and low as specified below:

High: Scores above mean + standard deviation ($> \overline{X} + \delta$)

Medium: Scores in between mean ± standard deviation ($\overline{X} \pm \delta$)

Low: Scores below mean - standard deviation ($< \overline{X} - \delta$)

(d) Correlation coefficient

In order to know the relationship between any two variables, correlation coefficient was used. Correlation coefficients between dependent variable and selected independent variables were calculated by using the following formula:

$$r = \frac{\sum XiYi - \frac{\left(\sum Xi\right)\left(\sum Yi\right)}{N}}{\sqrt{\sum Xi^2 - \frac{\left(\sum Xi\right)^2}{N}}\sqrt{\sum Yi^2 - \frac{\left(\sum Yi\right)^2}{N}}}$$

where,

r = Correlation coefficient

N = Number of paired observations

Xi = Value of X variable for ith pair

Yi = Value of Y variable for ith pair

The significance of correlation coefficient was tested by using following formula:

$$t = r\sqrt{\frac{N-2}{1-r^2}}$$

d.f. = N-2

The value of r always lies between -1 to +1. Positive value of r indicates a tendency of X and Y to increase or decrease together, negative values of r indicates a tendency of increasing one variable with the decrease of other variable and vice-versa. The 'r' table was used for obtaining significance/non-significance of the correlation coefficients.

(e) Regression Analysis

Regression is a measure of average relationship between the variables. With its help, we can know the average probable change in one due to the others. For the purpose of the study the relationship between dependent variable Y and the selected independent variables (X_1, X_2, X_3, X_4, X_5, X_6, X_7 and X_8) was obtained by fitting the multiple linear regression equation as follows:

$$Y = a + b_1X_1 + b_2X_2 + b_3X_3 + b_4X_4 + b_5X_5 + b_6X_6 + b_7X_7 + b_8X_8$$

where,

Y = Predicted value of the dependent variable.

a = Intercept constant, calculated in the following manner.

$$a = \overline{Y} - b_1\overline{X}_1 - b_2\overline{X}_2 - b_3\overline{X}_3 - b_4\overline{X}_4 - b_5\overline{X}_5 - b_6\overline{X}_6 - b_7\overline{X}_7 - b_8\overline{X}_8$$

b_i = Partial regression coefficient, which represents the amount of change in 'Y' that can be associated with a unit change in any one of the X_i when the remaining independent variables were fixed. The significance of partial regression coefficient can be tested by t test with n-2 d.f. as

$$t = \frac{bi}{SE(bi)}$$

SE (bi) = Standard error of partial regression coefficient bi.

(f) Multiple Correlation coefficients

Multiple correlation coefficient (R) was calculated by the formula

$$R = \sqrt{\frac{\text{Regression S.S.}}{\text{S.S.(Y)}}}$$

Regression S.S. = $b_1S_P(X_1Y) + b_2S_p(X_2Y) + b_3S_p(X_3Y) + b_4S_p(X_4Y) + b_5S_p(X_5Y)$

$$S.S. = \sum Yi^2 - \frac{\left(\sum Yi\right)^2}{N}$$

$$S_P(XiYi) = \sum XiYi - \frac{\left(\sum Xi\right)\left(\sum Yi\right)}{N}$$

i = 1, 2, 3,. N

Significance of multiple correlation coefficient (R) was tested by using 'F' test formula

$$F(K, N\text{-}K\text{-}1) = \frac{R^2}{1-R^2} \frac{N-K-1}{K}$$

d.f. = K, N-K-1

where,

K = Number of independent variables

N = No. of respondents in the sample

(g) Cobb-Douglas Production Function or Power Function

It was used in the following form:

$$Y = aX_1^{b_1} X_2^{b_2} X_3^{b_3} X_4^{b_4} X_5^{b_5} X_6^{b_6} X_7^{b_7} X_8^{b_8}$$

where,

Y = Output (Communication behaviour)

a = Constant

X_1 = Size of land holding

X_2 = Annual income

X_3 = Interpersonal interaction

X_4 = Innovation proneness

X_5 = Value orientation

X_6 = Achievement motivation

X_7 = Socio-economic status

X_8 = Education.

b_1, b_2, b_3, b_4, b_5, b_6, b_7 and b_8 are regression coefficients.

Chapter 4

Findings and Discussion

This chapter presents, the entire findings of investigation along with discussion which have been arrived after subjecting data to statistical analysis and interpretation, keeping in view, the objectives of the study. For the convenience of the presentation, the findings and discussion have been grouped under the following sections:

1. To study the communication behaviour of the farmers towards farm education programmes of television.
2. To find out the preference of television in comparison to other agricultural information sources.
3. To examine the association between selected independent variables namely socio-economic status, education, size of land holding, annual income, interpersonal interaction, innovation proneness, value orientation and achievement motivation with the communication behaviour of the farmers towards farm education programmes of television.
4. To identify the agricultural technology for farm education programmes of television using entertainment farm education format and ranking of programmes if available from television centres and respondents.
5. To develop an appropriate model for improving the communication behaviour of farmers.

Objective 1: To Study the Communication Behaviour of the Farmers Towards Farm Education Programmes of Television

(A) Information Input Behaviour

The respondents were asked to indicate the sources by which they did update themselves with the scientific information.

Table 3: Distribution of Respondents on their Frequency of using different Sources of Information

Sl.No.	*Sources of Information*	*Frequency of Use of Different Sources of Information*			*Ranks*	
		Often	*Occasionally*	*Never*	*Often*	*Never*
1.	Extension personnel of KVK	6 (4.00)	9 (6.00)	135 (90.00)	X	I
2.	Extension personnel of C.D. Blocks	72 (48.00)	63 (42.00)	15 (10.00)	II	IX
3.	Salesmen of Agril. inputs	70 (46.66)	52 (34.66)	28 (18.66)	III	VIII
4.	Local leaders	35 (25.33)	50 (33.33)	65 (43.38)	VIII	III
5.	Progressive farmers	60 (40.00)	35 (23.32)	55 (36.66)	IV	IV
6.	T.V.	87 (58.00)	63 (44.00)	00 (00.00)	I	X
7.	Radio	57 (38.00)	64 (42.66)	29 (19.33)	V	VII
8.	Extension Publications	23 (15.33)	57 (44.66)	60 (45.30)	IX	II
9.	Neighbours	41 (24.33)	67 (44.66)	42 (28.00)	VII	VI
10.	Relatives and friends	43 (28.66)	60 (40.00)	47 (31.33)	VI	V

* Figures in parentheses indicate percentages.

The above Table shows that the farmers oftenly get the information from T.V. (58.00 per cent), C.D. Block (48.00 per cent), salesmen of agril. inputs (46.66 per cent), progressive farmers (40.00 per cent), radio (38.00 per cent), relatives and friends (28.66 per cent), neighbors (27.33 per cent), local leaders (25.33 per cent), extension publications (16.66 per cent) and extension personnel of KVK (4.00 per cent) respectively.

The farmers also got the information occasionally from neighbours (44.66 per cent), extension publications (44.66 per cent), T.V. (44.00 per cent), radio (42.66 per cent), extension personnel of C.D. Blocks (42.00 per cent), relatives and friends (40.00 per cent), salesmen of agril. inputs (34.66 per cent), local leaders (33.33 per cent), progressive farmers (23.32 per cent) and extension personnel of KVK (6.00 per cent) respectively.

The farmers who never got the information from extension personnel of KVK (90.00 per cent), extension publications (45.30 per cent), local leaders (43.38 per cent), progressive farmers (36.66 per cent), relatives and friends (31.33 per cent), neighbours (28.00 per cent) radio (19.33 per cent), salesmen of agril. inputs (18.66 per cent), extension personnel of C.D. Bocks (10.00 per cent) and T.V. (00.00 per cent) respectively.

In case of oftenly seeking information sources T.V. was ranked first. It might be due to fact that T.V. was used oftenly for entertainment- cum-education purposes. Extension personnel were ranked second. This could be due to the fact that extension personnel were efficient in technology dissemination and were oftenly going to the villages. Salesmen of agril. information were ranked third. It could be due to fact that the credibility of salesmen is increasing day-by-day. Progressive farmers were ranked fourth. It might be due to fact that the farmers were constantly having contact with the progressive farmers for getting the information. Radio was ranked fifth in case of oftenly seeking information. It could be due to fact that radio as entertainment-cum-education source is very popular in India. Apart from these sources, the other sources like relatives and friends (II), neighbours (VII), local leaders (VIII), extension publications (IX) and extension personnel of KVK (X) were oftenly preferred by the farmers.

In case of never getting information sources, extension personnel of KVK were ranked first. It could be due to fact that KVKs were far from the villages. Extension publications were ranked second. It might be due to fact that the farmers were not interested to get the information from the extension publications. Local leaders were ranked third. It might be due to fact that local leaders were having differences with the common farmers. Progressive farmers were ranked fourth. It could be due to fact that progressive farmers were having contact with the limited number of farmers of the villages. Apart from these sources the farmers never got the information from relatives and friends (V), neighbours (VI), radio (VII), salesmen of agril. inputs (VIII), extension personnel of C.D. Blocks (IX) and T.V. (X) respectively.

The results are in accordance with the results of Sinha and Prasad (1966), Williams (1969), Ambastha (1974), Bhangoo and Kawer (1994) and Singh and Singh (1997).

(B) Information Processing Behaviour of Respondents

(a) Information Evaluation

It is clear from the Table 4 that respondents had evaluated the information oftenly by discussing with progressive farmers (66.00 per cent), local leaders (58.00 per cent), neighbours (38.66 per cent), elder family members (27.23 per cent), thinking about technical feasibility (26.66 per cent), on the basis of their past experiences (6.00 per cent) and by discussing with SHG/farm association (2.66 per cent).

The respondents had evaluated the information occasionally by thinking about technical feasibility (20.00 per cent), discussing with elder family members (18.00 per cent), by discussing with SHG/farm association (14.00 per cent), progressive farmers (12.00 per cent), local leaders (6.66 per cent), neighbours (6.00 per cent), on the basis of their past experiences (5.33 per cent). The percentages of respondents who never evaluated the information by these methods were 88.66, 60, 56.00, 54.00, 53.33, 35.33 and 22.00 respectively.

Table 4: Distribution of Respondents on the Basis of Information Evaluation, Information Storage and Information Transformation

(N=150)

Sl.No.	*Statements*	*Frequency*			*Ranks*	
		Often	*Occasionally*	*Never*	*Often*	*Never*
(a)	**Information evaluation**					
1.	Discuss with elder family members	41 (27.33)	28 (18.66)	81 (54.00)	IV	IV
2.	Discuss with neighbours	57 (38.00)	9 (6.00)	84 (56.00)	III	III
3.	Discuss with progressive farmers	99 (66.00)	18 (12.00)	33 (22.00)	I	VII
4.	Discuss with local leaders/ key communicators	87 (58.00)	10 (6.66)	53 (35.33)	II	VI
5.	Discuss in light of past experiences	9 (6.00)	8 (5.33)	133 (88.66)	VII	I
6.	Thinking about technical feasibility	40 (26.66)	30 (20.00)	80 (53.33)	V	V
7.	Discuss with SHGs/ farm association	39 (26.00)	21 (14.00)	90 (60.00)	VI	II
(b)	**Information storage**					
1.	By memorization	49 (32.66)	5 (3.33)	96 (64.00)	I	III
2.	Writing in general notebook	38 (25.33)	43 (28.66)	69 (46.00)	II	IV
3.	Preparing subjectwise files	7 (4.66)	4 (2.66)	139 (92.66)	IV	I
4.	By preserving the printed matter	12 (8.00)	6 (4.00)	132 (88.00)	III	II
(c)	**Information transformation**					
1.	Rearrange the important information as per farmers needs	69 (46.00)	30 (20.00)	51 (34.00)	I	II
2.	Rearrange the information in local dialect	20 (13.33)	19 (12.66)	111 (74.00)	II	I

* Figure in parentheses indicate percentages.

In case of oftenly information evaluation, discussion with progressive farmers was at first place. It might be due to fact that the farmers were having more authenticity on them as compared to the others. Discussion with the local leaders/ key communications was at second place. It could be due to fact that farmers were having faith on him and liked to discuss with him. Discussion with neighbours was at third place. It could be due to fact that due to proximity and authenticity farmers liked to discuss with them. Discussion with elder family members (IV), thinking about technical feasibility (V), discussion with SHGs/farm associations (VI) and discussion in light of past experiences (VII) were also used oftenly by the farmers for evaluation of information.

In never evaluation case, discussion in light of past experiences was at first rank. It could be due to fact that farmers were not too much interested to use their past experiences. SHGs/farm association was at second place. It might be due to fact that the farmers were not serious to discuss in their SHGs or their association. Discussion with neighbours was at third rank. It could be due to fact that they were not thinking good to discuss with their neighbours. While discussion with elder family members (IV), thinking about technical feasibility (V), local leaders/ key communicators (VI) and progressive farmers (VII) were in ranking for never information evaluation.

(b) Information Storage

The Table 4 further shows that the respondent's oftenly stored the information by memorization (32.66 per cent), writing in general notebooks (25.33 per cent), by preserving the printed matter (8.00 per cent) and preparing subject wise files (4.66 per cent) respectively. The percentages of respondents who use the information storage occasionally by these methods were 28.66, 4.00, 3.33, and 2.66 respectively. The percentage of respondents who never used the information storage by these methods were 92.66, 88.00, 64.00 and 46.00 respectively.

In case of oftenly storage of information by memorization was at first place. It could be due to fact that the farmers perceived the information properly and retained in their minds very easily. Writing in general notebook was at second place. It could be due to fact that farmers were very much interested in storing the information by writing them down. By preserving the information (III) and preparing subject wise files (IV) were also used oftenly by the farmers.

In never information storage case, preparing subject wise was at first rank. It could be due to fact that farmers were not interested to store the information in subject wise way. By preserving the printed matter was at second place. It means the farmers were not taking too much interest for storing the material through the printed matter. By memorization (III) and writing in general notebooks (IV) were in ranking for never storing the information.

(c) Information Transformation

It is clear from the Table that the respondent's oftenly transformed the information by rearranging the important information as per their needs (46.00 per cent) and rearranging the information in local dialect (13.13 per cent). The percentages of respondents who occasionally transformed the information were 20.00 and 12.66 respectively. The percentages of respondents who never transformed information by these methods were 74.00 and 34.00 respectively.

In case of oftenly information transformation, the farmers' rearrangement of the important information as per their needs was at first place. It could be due to fact that the farmers got the information unsystematic way. While the rearrangement of the information in local dialect was at second place. It could be due to fact that the farmers feel easy to repeat the information if transformed in local dialect.

In case of never transformation case, rearrangement the information in local dialect was at first place. It could be due to fact that farmers were not taking keen interest for rearrangement of information in local dialect while the rearrangement of the important information as per their needs was at second rank. It could be due to fact that farmers were not arranging the important information as per their needs.

The findings are in accordance with the findings of Akhoury (1973), Ambastha (1974) and Pandey (1979).

(c) Information Output Behaviour

Table 5: Distribution of Respondents on the Basis of Information Output Behaviour

Sl.No.	Statements	Frequency			Ranks	
		Often	Occasionally	Never	Often	Never
1.	To my family members	121 (80.66)	16 (10.66)	13 (8.66)	I	VIII
2.	To my relatives	26 (17.33)	41 (27.33)	83 (55.33)	V	IV
3.	To my neighbours	92 (61.33)	7(4.66)	51(34.00)	III	V
4.	To my friends	98 (65.33)	24 (16.00)	28 (18.66)	II	VII
5.	To the person who contacted me	21 (14.00)	39 (26.00)	90 (60.00)	VIII	II
6.	To all the persons known to me	25 (16.66)	41 (27.33)	84 (56.00)	VI	III
7.	To the farmers of neighbouring villages	22 (14.66)	16 (10.66)	110 (73.33)	VII	I
8.	To those who are cultivating in my land	91 (60.66)	15 (10.00)	44 (29.33)	IV	VI

* Figures in parentheses indicate percentages.

The farmers after getting the information and processing it, disseminate to others.

It is clear from the Table 5 that the farmers disseminated the information oftenly to their family members (80.66 per cent), friends (65.33 per cent), neighbours (61.33 per cent), those who cultivate in their land (60.66 per cent), relatives (17.33 per cent), the persons who were known to him (16.66 per cent), to the farmers of neighbouring villages (14.66 per cent) and the other persons who contacted him (14 per cent) respectively. The percentages of farmers who disseminated the information occasionally to others were 27.33, 27.33, 26, 16, 10.66, 10 and 4.66 respectively. The percentage farmers who never disseminate the information to others were 73.33, 60, 56, 55.33, 34, 28.66, 18.66 and 8.66 respectively.

In case of oftenly information output behaviour, family members were at first place. It could be the proximity of the farmers to their family members. Friends were at second place. It could be due to the fact that friends were in constant touch and had proximity with the farmers. Neighbours were at third place. It could be the proximity and nearness of the farmers with their neighbours. Cultivators in farmers land were at fourth rank. It could be due to the fact that farmers were constantly disseminating the new technologies to the persons who were cultivating in their lands. Relatives (V), all the persons who were known to the farmer (VI), the farmers of neighbouring villages (VII) and the persons who contacted the farmer (VIII) were also used oftenly for dissemination of information.

In never case, farmers of neighbouring villages were at first rank. It could be due to fact that farmers of one village were not in constant contact with the farmers of other villages. The persons who contacted the farmer were at second rank. It could be due to fact that the farmers were not interested generally to tell about the information to the person who were in contact with him. The persons who were known to him were at third rank. it could be due to fact that the farmers were not showing any type of interest to tell them about the new information. Relatives (IV), neighbours (V), those who were cultivating in his land (VI), friends (VII) and family members (VIII), were never able to get the information. The findings are in line with Sunderswamy (1971), Singh and Singh (1977) and Pandey (1979).

Table 6: Distribution of Respondents According to their Communication Behaviour Towards Farm Education Programmes of Television

(N=150)

Sl.No.	Level of Communication Behaviour	Frequency of Respondents
1	Low (Below $\overline{X} - SD$)	24 (16.00)
2	Medium (in between $\overline{X} \pm SD$)	83 (55.33)
3	High (more than $\overline{X} + SD$)	43 (28.66)

* Figures in parentheses indicate the percentages.

It is clear from the above Table that 16.00 per cent respondents had low communication behaviour towards farm education programmes of television.55.33 per cent respondents had medium communication behaviour towards farm education programmes of television and 28.66 per cent respondents had high communication behaviour towards farm education programmes of television.

The findings of the study highlighted that majority of farmers have medium and high communication behaviour towards farm education programmes of T.V. It shows that farmers of the study area are progressive in nature. This could be due to the fact that farmers knew the utility of the television and they realized that television was of great help to them in increasing the food production. The finding is in line with Rajput (1993), Singh and Lal (1997) and Singh and Singh (1997).

Objective 2: To Find Out the Preference of Television in Comparison to other Agricultural Information Sources

An attempt was made to know the preference of television as compared to other sources of agricultural information along with T.V. as a source. For this purpose a list of all the possible sources was prepared and responses were recorded on a three point continuum ranging from highly preferred to least preferred. Mean scores for each were calculated and sources were ranked according to their respective mean scores. The results are presented in the Table 7.

Table 7: Preference of Television in Comparison to other Agricultural Information Sources Along with Television as a Source

Sl.No.	*Sources*	*Highly Preferred (Frequency)*	*Preferred (Frequency)*	*Least Preferred (Frequency)*	*Mean Score*	*Ranks*
1.	Extension personnel of KVK	19	37	104	1.56	X
2.	Extension personnel of C.D. Blocks	80	51	19	2.40	I
3.	Salesmen of agril. inputs	67	50	33	2.29	II
4.	Local leaders	47	47	56	1.92	VIII
5.	Progressive farmers	57	56	37	2.14	V
6.	T.V.	55	69	26	2.19	III
7.	Radio	59	59	32	2.18	IV
8.	Extension publications	22	49	79	1.62	IX
9.	Neighbours	43	65	42	2.00	VI
10.	Relatives and friends	42	62	46	1.97	VII

The figures in Table 7 reveal that extension personnel of C.D. Blocks were ranked first in order to preference. This could be due to the fact that extension personnel of C.D. Blocks were educated, efficient, trained and their services were prompt that made the farmers to approach and utilize them. On the other hand salesmen of agril. inputs were preferred by the farmers with second priority. This could be due to fact that the faith of farmers in the salesmen of agril. inputs is increasing due to their prompt services. T.V. was ranked third in order of preference. This could be due to fact that farmers might have taken it for entertainment-cum-educational purpose. Radio came up as the fourth important source, which was preferred by the farmers. This could be due to fact that radio as a means of entertainment cum-education has become very popular during the last few decades in India. Progressive farmers were ranked fifth priority. This could be due to fact that farmers of the area might go to the progressive farmers for advice and information and had good contact with them. Apart from these sources, the other sources like neighbours, relatives and friends, local leaders, extension, publications and extension personnel of KVK were least preferred by the farmers.

The finding is in line with the findings of Singh and Singh (1971), Singh and Sharma (1973), Mehra (1976), Chandra and Cherian (1991), Khan and Paracha (1994), Chauhan (1997) and Lal (2002).

Objective 3: To Examine the Association between Selected Independent Variables Namely Socio-Economic Status, Education, Size of Land Holding, Annual Income, Interpersonal Interaction, Innovation Proneness, Value Orientation and Achievement Motivation with the Communication Behaviour of the Farmers towards Farm Education Programmes of Television

Table 8: Correlation between Selected Iindependent Variables with the Communication Behaviour (Dependent variable) of the Farmers

Sl.No.	*Selected Independent Variables*		*Values of Correlation Coefficients (r)*
1	Socio-economic status	(X_1)	0.4225**
2.	Education	(X_2)	0.7947**
3.	Size of land holding	(X_3)	0.2648**
4.	Annual income	(X_4)	0.0086
5.	Interpersonal interaction	(X_5)	0.1942*
6.	Innovation proneness	(X_6)	0.2140**
7.	Value orientation	(X_7)	0.1847*
8.	Achievement motivation	(X_8)	0.4375**

** Significant at 0.01 level of significance.

* Significant at 0.05 level of significance.

A critical analysis of the data presented in Table 8, reveals that socio-economic status, education, size of land holding, innovation proneness and achievement motivation were positively and significantly correlated with the communication beahviour of the farmers at 1 per cent level of significance while interpersonal interaction and value orientation were positively and significantly correlated with the communication behaviour of the farmers at 5 per cent level of significance. It means that these variables exerted a significant effect on the communication behaviour of the farmers towards farm education programmes of television.

This rejects the hypothesis H_{01}, H_{02}, H_{03}, H_{05}, H_{06}, H_{07} and H_{08} i.e. there is no association between the communication behaviour of the farmers towards farm education programmes of television and their socio-economic status, education, land holding, interpersonal interaction innovation proneness, value orientation and achievement motivation. The alternate hypothesis is accepted that is there is relationship between socio-economic status, education, land, interpersonal interaction, innovation proneness, value orientation and achievement motivation with the communication behaviour of the farmers. Annual income was non-significant for developing the communication behaviour of the farmers towards farm education programmes of television. This accepts the hypothesis (H_{04}).

The socio- economic status of the farmers was positively and significantly associated with the communication behaviour of farmers towards farm education programmes of T.V. Hence, hypothesis (H_{01}) stated in null form that "There is no association between socio-economic status and the communication behaviour of the farmers towards farm education programmes of T.V" was rejected. It means socio-economic status of the farmers' exerted highly significant influence on the communication behaviour of the farmers towards farm education programmes of T.V.

This might be due to fact that innovative farmers with high socio-economic status pay much attention to the farm education programmes of T.V. to keep abreast with the latest knowledge so as to increase their agricultural production.

The results are accordance with the results of Ambastha (1974), Pandey (1979), Kaur (1982) and Singh *et al.* (1994).

Level of education was positively and significantly associated with the communication behaviour of the farmers towards farm education programmes of T.V. Hence, the hypothesis (H_{02}) stated in null form that "There is no association between level of education and the communication behaviour of the farmers towards farm education programmes of T.V." was rejected. It means that farmers with higher education were motivated to pay more response towards farm education programmes of T.V.

The results are in accordance with the findings of Ambastha (1974), Pandey (1979), Singh and Gupta (1979), Yadav (1983) and Singh and Singh (1997) who also found that communication behaviour of the farmers was positively and significantly associated with their level of education.

Size of land holding was positively and significantly associated with the communication behaviour of the farmers towards farm education programmes of T.V. Hence, the hypothesis (H_{03}) stated in null form that "There is no association between the size of land holding and the communication behaviour of the farmers towards farm education programmes of T.V." was rejected. It means the size of land holding of farmers exerted highly significant influence on their communication behaviour towards farm education programmes of T.V.

This might be due to fact that information about new agril. technology was equally important for the farmers in respect of their farm size.

The results are in accordance with the results of Singh *et al.* (1992), Singh *et al.* (1994) and Singh and Singh (1997).

Annual income of the farmers was found to be non-significantly associated with the communication behaviour of farmers towards farm education programmes of T.V. Hence, the hypothesis stated in null form that "There is no association between the income and communication behaviour of the farmers towards farm education programmes of T.V." was accepted. It means the income could not exert a significant influence on the communication behaviour of the farmers towards farm education programmes of T.V.

The results seem to be quite natural due to the fact that income factor is not a barrier to seeking agricultural information through T.V. and people of even lower income could receive the agricultural information through T.V.

The results are in accordance with the results of Singh *et al.* (1994).

Interpersonal interaction was found to be positively and significantly associated with the communication behaviour of the farmers towards farm education programmes of T.V. Hence, the hypothesis (H_{05}) stated in null form that "There is no association between interpersonal interaction and the communication behaviour towards farm education programmes of T.V." was rejected. It means the interpersonal interaction of the farmers' exerted highly significant influence on their communication behaviour towards farm education programmes of T.V.

This finding is in line with Barbato and Perey (1992).

Interaction between and among the individuals in a face-to-face situation was more effective. Hence, its significant relationship has been established.

Innovation proneness was found to be positively and significantly associated with the communication behaviour of the farmers towards farm education programmes of T.V. Hence, the hypothesis (H_{06}) stated in null form that "There is no association between innovation proneness and the communication behaviour of the farmers towards farm education programmes of T.V." was rejected. It means the innovation proneness of the farmer exerted highly significant influence on the communication behaviour of the farmers towards farm education programmes of T.V. The finding is in line with Singh and Prasad (1974) and Pandey (1979).

This might be due to fact that they are satisfied with the present status of living.

Value orientation was positively and significantly associated with the communication behaviour of the farmers towards farm education programmes of T.V. Hence, the hypothesis (H_{07}) stated in null form that "There is no association between value orientation and communication behaviour of the farmers towards farm education programmes of T.V." was rejected. Hence, value orientation exerted significant influence on the communication behaviour of the farmers towards farm education programmes of T.V.

Achievement motivation was found positively and significantly associated with the communication behaviour of the farmers towards farm education programmes of T.V. Hence, the hypothesis (H_{08}) stated in null form that "There is no association between achievement motivation and communication behaviour of the farmers towards farm education programmes of T.V." was rejected. Hence, achievement motivation exerted significant influence on the communication behaviour of the farmers towards farm education programme of T.V.

To predict the important independent variables the technique of multiple regression was used. The technique was used to determine the effect of these selected variables on the dependent variable namely communication behaviour of the farmers towards farm education programmes of T.V. The following symbols have been used to denote the variables.

Dependent variable: Y = Communication behaviour

Independent variables : X_1 - socio-economic status, X_2 - Education, X_3 - Land,

X_4 - Annual Income, X_5 - Interpersonal interaction, X_6 - Innovation proneness, X_7 - Value orientation, X_8 - Achievement motivation.

Table 9: Multiple Regression (Linear) Analysis of the Independent Variables with the Communication Behaviour (Dependent variable) of the Farmers

Sl.No.	*Independent Variables*		*Regression Coefficient 'b'*	*S.E. of b*	*'t'*
1.	Socio-economic status	(X_1)	0.0945	0.0387	2.439*
2.	Education	(X_2)	6.0328	0.4790	12.593**
3.	Size of land holding	(X_3)	0.0364	0.0135	2.677**
4.	Annual income	(X_4)	5.13178	3.3784	1.519
5.	Interpersonal interaction	(X_5)	0.1862	0.6858	2.272*
6.	Innovation proneness	(X_6)	2.3817	1.3284	1.7928
7.	Value orientation	(X_7)	0.0793	0.1610	0.492
8.	Achievement motivation	(X_8)	0.1423	0.1854	0.768

R^2 = 0.6564, Intercept constant (a) = 82.93.

'F' value = 33.67**; at d.f. 8, 141.

* Significant at 0.05 level of significance.

**Significant at 0.01 level of significance.

A close study of the data in Table 9 indicates that all the eight independent variables taken together explained to the extent of 65.64 per cent variation in the communication behaviour of farmers towards farm education programmes of T.V. The calculated 'F' value was 33.67** at 8 and 141 degree of freedom which was significant at 0.01 level of significance. Thus, the result implied that all eight independent variables taken together would account for a significant amount of variation in the communication behaviour of the farmers towards farm education programmes of T.V.

Further, the 't' test of significance expressed that coefficient of regression ('b' value) was found non-significant for annual income, innovation proneness, value orientation and achievement motivation which means that these variables were not contributing significantly in predicting the communication behaviour of the farmers towards farm education programmes of T.V. On the other hand coefficient of regression was found positively significant for socio-economic status and interpersonal interaction at 0.05 level of significance while education and size of land holding were found positively significant at 0.01 level of significance. It means that these variables were contributing significantly in predicting the communication behaviour of the farmers towards farm education programmes of T.V.

It has been adequately observed that socio-economic status, education, land and interpersonal interaction are important variables in the sphere of communication behaviour of the farmers towards farm education programmes of T.V. Therefore, attention has to be made with due care in socio-economic status, education, land and interpersonal interaction characteristics for effective communication behaviour of the farmers towards farm education programmes of T.V. Conclusion can be drawn that socio-economic status, education, land and interpersonal interaction are the most contributing factors towards farm education programmes of T.V.

Objective 4: To Identify the Agricultural Technology for Farm Education Programmes of Television using Entertainment Farm Education Format and Ranking of Programmes if Available from T.V. Centres and Respondents

Out of various agricultural technologies the following technologies were identified using entertainment farm education format.

Table 10: Identification of Agricultural Technology for Farm Education Programmes of Television using Entertainment Farm Education Format

Sl.No.	*Agricultural Technology*	*Frequency*
1.	Methods of using insecticides and pesticides	83 (55.33)
2.	Guidance about dairy	37 (24.66)
3	Guidance about poultry production	19 (12.66)
4.	Techniques of improving soil conditions	11 (7.33)

* Figures in parentheses indicate the percentages.

It is clear from the above Table that 55.33 per cent farmers liked the methods of using insecticides and pesticides to be telecasted through entertainment. 24.66 per cent liked the guidance about dairy to be telecasted through entertainment. 12.66 per cent liked the guidance about poultry production to be telecasted through entertainment. While 7.33 per cent said that they liked the techniques of improving soil conditions to be telecasted through entertainment.

Table 11: Ranking of Programmes

Sl.No.	*Programmes*	*Frequency*	*Rank*
1.	Krishi Darshan	139 (92.66)	I
2.	Mera Pind Mere Khet	7 (4.66)	II
3.	Sadi Dharti Sade Lok	4 (2.66)	III

* Figures in parentheses indicate the percentages.

It is clear from above Table that 92.66 per cent farmers liked to watch Krishi Darshan. That is why it is ranked I while Mera Pind Mere Khet was watched by 4.66 per cent farmers and ranked II. Sadi Dharti Sade Lok was watched by 2.66 per cent farmers and ranked III.

It is clear from Table 12 that 86 per cent of the respondents said that they liked mostly the methods of using insecticides and pesticides which ranked I. Guidance of dairy was liked by 78 per cent of respondents and ranked II. While guidance about animal husbandry ranked III. Monthly schedule of farming practices of different local crops ranked IV while information relating to institution providing financial support to the farmers is ranked V. Techniques of improving soil conditions is ranked VI while guidance about poultry production ranked VII.

Table 12: Ranking of Programmes Contents

Sl.No.	Items (Programme Contents)	Percentage of Farmers	Rank
1.	Methods of using insecticides and pesticides	86.00	I
2.	Techniques of improving soil conditions	57.00	VI
3.	Guidance about dairy	78.00	II
4.	Guidance about animal husbandry	73.00	III
5.	Guidance about poultry production	47.00	VII
6.	Monthly schedule of farming practices of different local crops	69.00	IV
7.	Information relating to institution providing financial support to the farmers	58.00	V

Objective 5: To Develop an Appropriate Model for Improving the Communication Behaviour of Farmers

In order to determine an appropriate model for the communication behaviour of farmers, firstly the linear function was fitted and the value of R^2 was found to be 0.6564 which was not very high. Then, a number of models were tried and the Cobb-Douglas model was considered as the best fit. The value of R^2 in this case was very high (0.9835). The Cobb-Douglas function is defined as:

$$Y = a\, X_1^{b_1}\, X_2^{b_2} \ldots\ldots\ldots\ \ldots\ldots X_n^{b_n}$$

where,

'Y' is the dependent variable,

'a' is the intercept constant,

X_i's are the independent variables and

b_i's are the regression coefficients.

The Cobb-Douglas equation has been worked out in the following way between the dependent variable (communication behaviour) and independent variables X_1, X_2, X_3, X_4, X_5, X_6, X_7 and X_8 (size of land holding, annual income, interpersonal interaction, innovation proneness, value orientation, achievement motivation, socio-economic status and education respectively).

Communication behaviour model is expressed as:

$$Y = 11.20\, X_1^{0.0748^*} X_2^{0.0982^{**}} X_3^{0.0330^*} X_4^{0.0134^*} X_5^{0.0608^*} X_6^{0.0212^{**}} X_7^{0.3059^{**}} X_8^{0.2906^{**}}$$

A close study of the data in Table 13 indicates that all the eight independent variables taken together explained to the extent of 98.35 per cent variation in the communication behaviour of farmers towards farm education programmes of T.V. It is clear that annual income, achievement motivation, socio-economic status and education were found to be positively significant at 1per cent level of significance while size of land holding, interpersonal interaction, innovation proneness and value orientation were found to be positively significant at 5 per cent level of significance. It means that these variables were important in predicting the communication behaviour of the farmers towards farm education programmes of television.

Table 13: Communication Behaviour and its Determinants - Multiple Regression Analysis

Item No.	*Variables*		*b*	*t-value*
1	Size of land holding	(X_1)	0.0748	2.2697*
2	Annual income	(X_2)	0.0982	2.9826**
3	Interpersonal interaction	(X_3)	0.0330	2.0972*
4	Innovation proneness	(X_4)	0.0134	2.2734*
5	Value orientation	(X_5)	0.0608	1.9827*
6	Achievement motivation	(X_6)	0.0121	3.5720**
7	Socio-economic status	(X_7)	0.3059	2.8723**
8	Education	(X_8)	0.2906	2.8647**

$R^2 = 0.9835$; a = 11.20

*Significant at 5 per cent level of significance.

** Significant at 1 per cent level of significance.

The significant coefficient of size of land holding with positive sign indicates that 1 per cent increase in the size of land holding would bring about an increase in the communication behaviour of the farmers towards farm education programmes of television by 0.0748 per cent by keeping other factors constant at their geometric mean level.

Similarly the significant coefficient of annual income with positive sign indicates that 1 per cent increase in annual income would bring about an increase in the communication behaviour of farmers towards farm education programmes of television by 0.0982 per cent by keeping other factors constant at their geometric mean level8 per cent by keeping other factors constant at their geometric mean level. ers towards farm eduction prog. Significant coefficient of interpersonal interaction would bring about an increase in the communication behaviour of the farmers towards farm education programmes of television by 0.0330 per cent by keeping other factors constant at their geometric mean level. The significant coefficient of innovation proneness with positive sign indicates that 1 per cent increase in innovation proneness would bring about an increase in the communication behavior of the farmers towards farm education programmes of television by 0.0134 per cent by keeping other factors constant at their geometric mean level. The significant coefficient of value orientation with positive sign indicates that 1 per cent increase in value orientation would bring about an increase in the communication behaviour of the farmers towards farm education programmes of television by 0.0608 per cent by keeping other variables constant at their geometric mean level. The significant coefficient of achievement motivation with positive sign indicates that 1 per cent increase in achievement motivation would bring about an increase in the communication behaviour of the farmers towards farm education programmes of television by 0.0121 per cent by keeping other factors constant at their geometric mean level. The significant coefficient of socio-economic status with positive sign indicates that 1 per cent increase in socio-economic status would bring about an increase in the communication behaviour of the farmers towards farm education programmes of television by 0.3059 per cent by keeping other factors constant at their geometric mean level. Similarly the significant coefficient of education with positive sign indicates that 1 per cent increase in education would bring about an increase in the communication behavior of the farmers towards farm education programmes of television by 0.2906 per cent by keeping other factors constant at their geometric mean level.

Also it is clear from the Table that summation of ($\Sigma bi = 0.8979$) which is less than one it means that it is decreasing returns to scale indicating that the rate of change in dependent variable would be less than the rate of change in independent variables. It signifies that increase in input by 1 per cent, would increase in output (communication behaviour) by less than 1 per cent.

Further from Figure 2, it is revealed that with the increase in size of land holding, the communication behavior of the farmers towards farm education programmes of television also increased but with the decreasing ratio. With the increase in size of land holding, communication behaviour also increased but ultimately a saturation point reached after that the communication behaviour of the farmers did not increase.

From Figure 3, it can be expressed that with the increase in annual income of the farmers the communication behaviour of the farmers towards farm education programmes of television also increased but with the decreasing ratio. Ultimately a saturation point reached after that the communication behavior of the farmers did not increase.

From Figure 4, it can be revealed that with the increase in interpersonal interaction, the communication behavior of the farmers towards farm education programmes of television increased with the decreasing ratio. With the increase in interpersonal interaction, ultimately a saturation point reached after that the communication behaviour of the farmers did not increase with the increase in interpersonal interaction.

From Figure 5, it is revealed that with the increase in innovation proneness, the communication behavior of the farmers towards farm education programmes of television increased with the decreasing ratio. Ultimately a saturation point reached after that the communication behavior of the farmers did not increase.

From Figure 6, it is revealed that the increase in value orientation, the communication behaviour of the farmers towards farm education programmes of television also increased but with the decreasing ration. With the increase in value

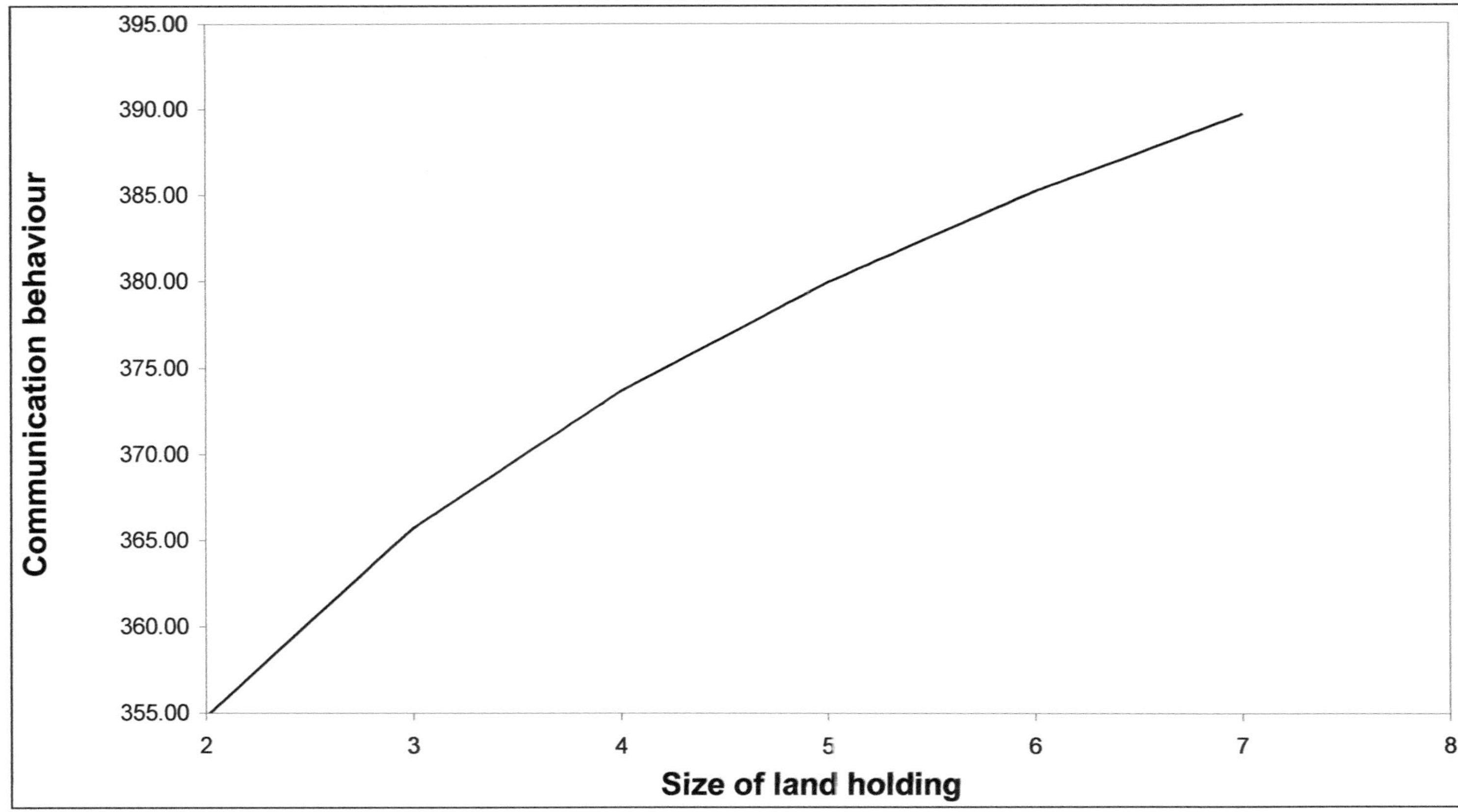

Figure 2: Exponential Model of Communication Behaviour (1).

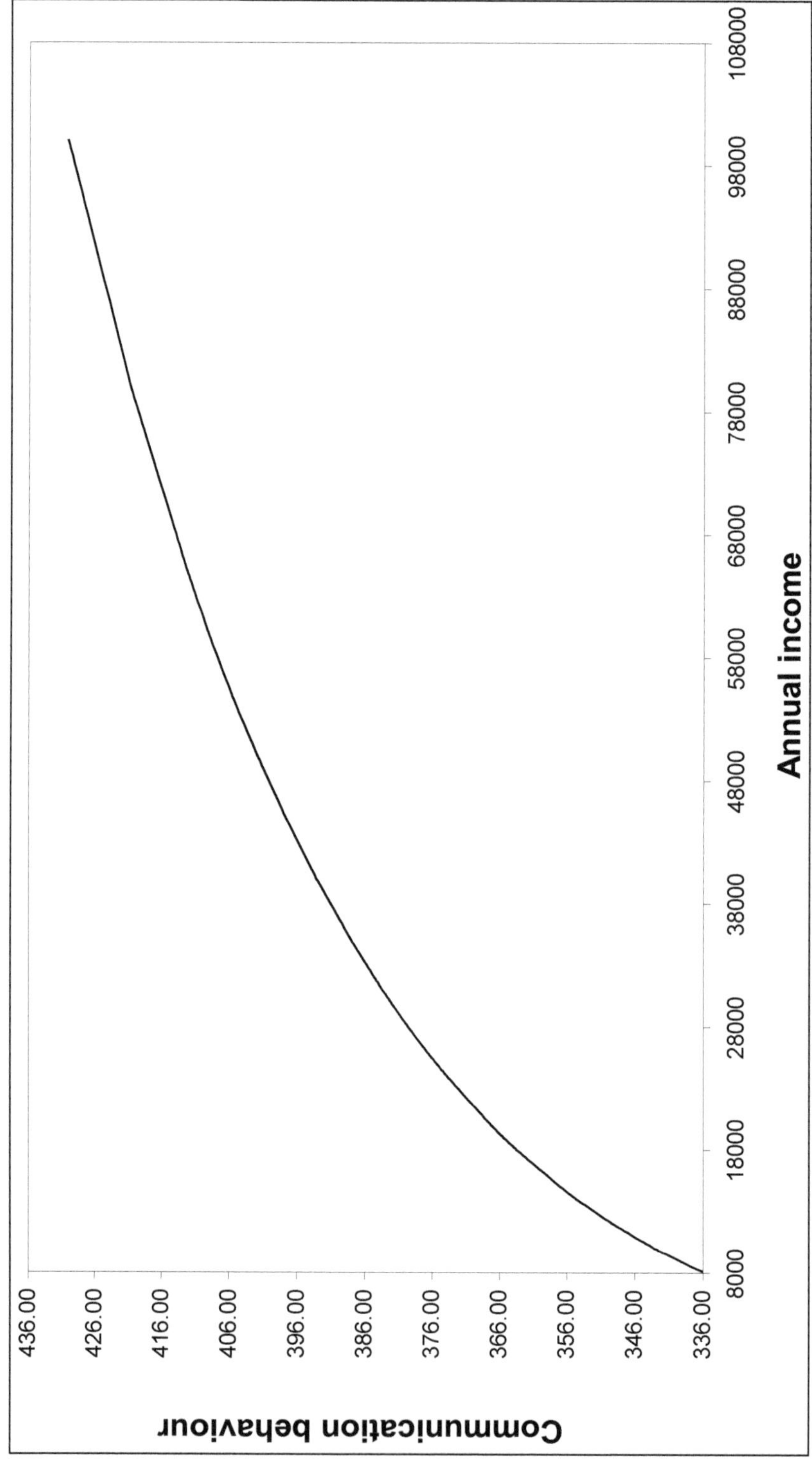

Figure 3: Exponential Model of Communication Behaviour (2).

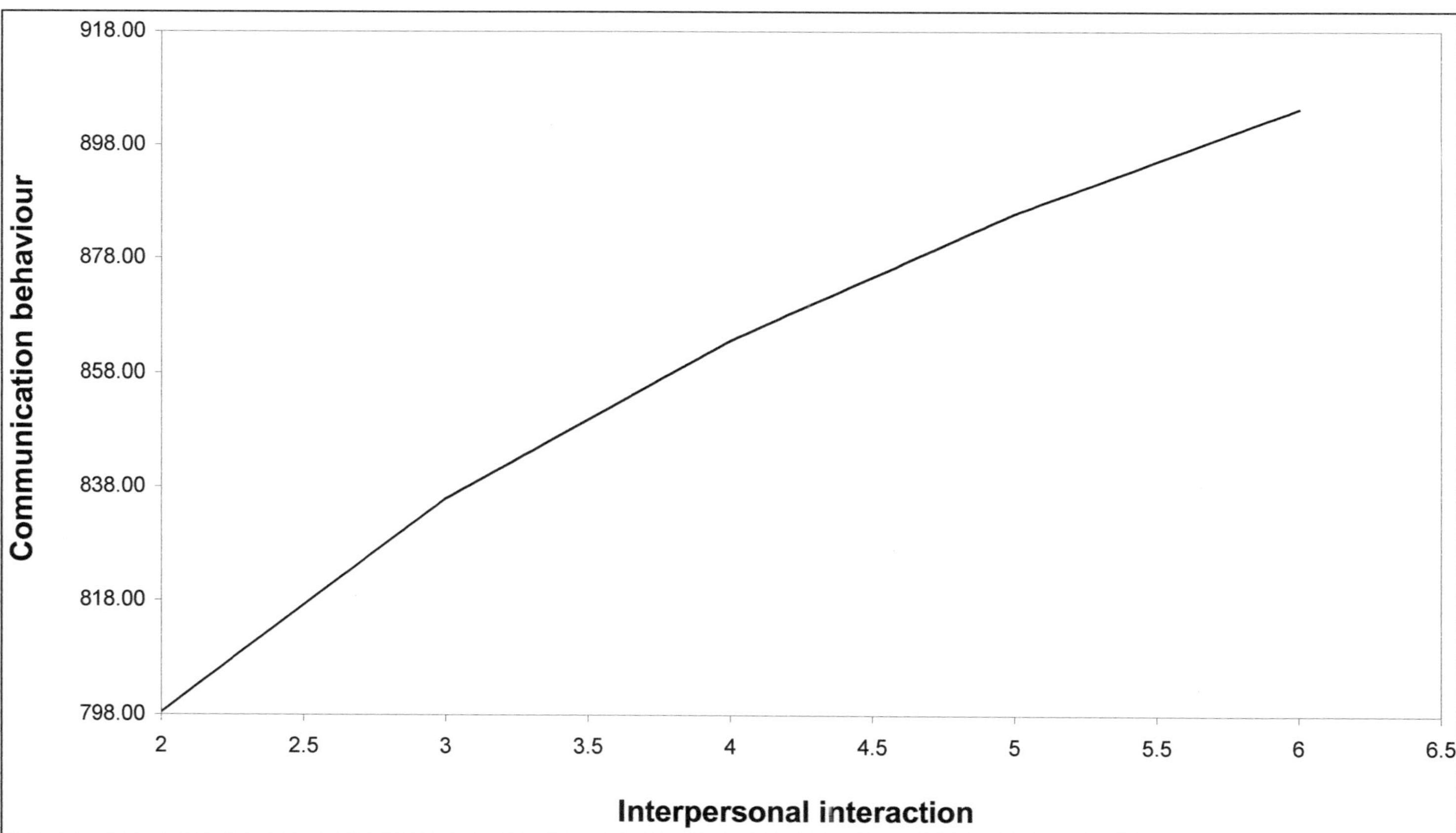

Figure 4: Exponential Model of Communication Behaviour (3).

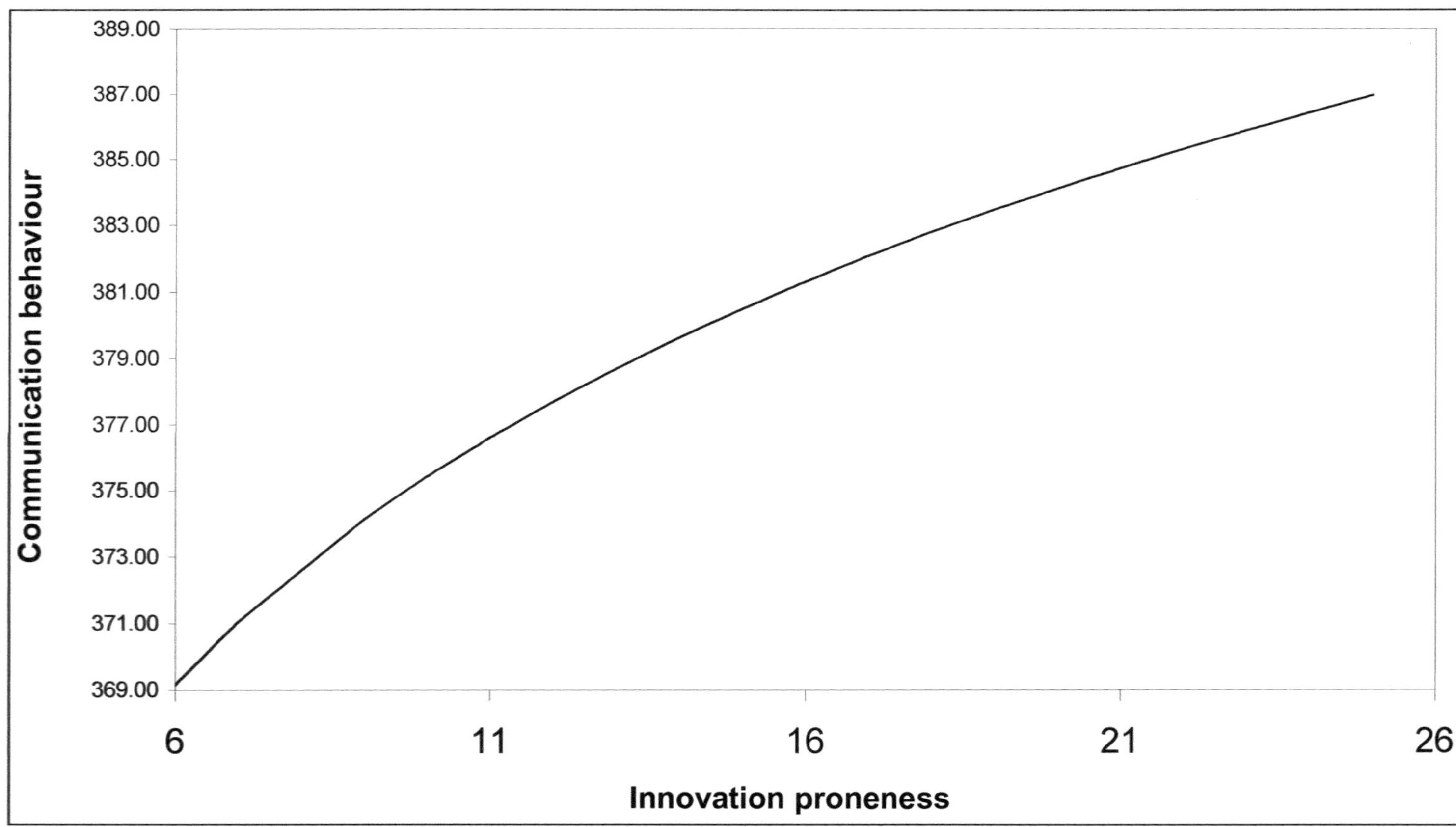

Figure 5: Exponential Model of Communication Behaviour (4).

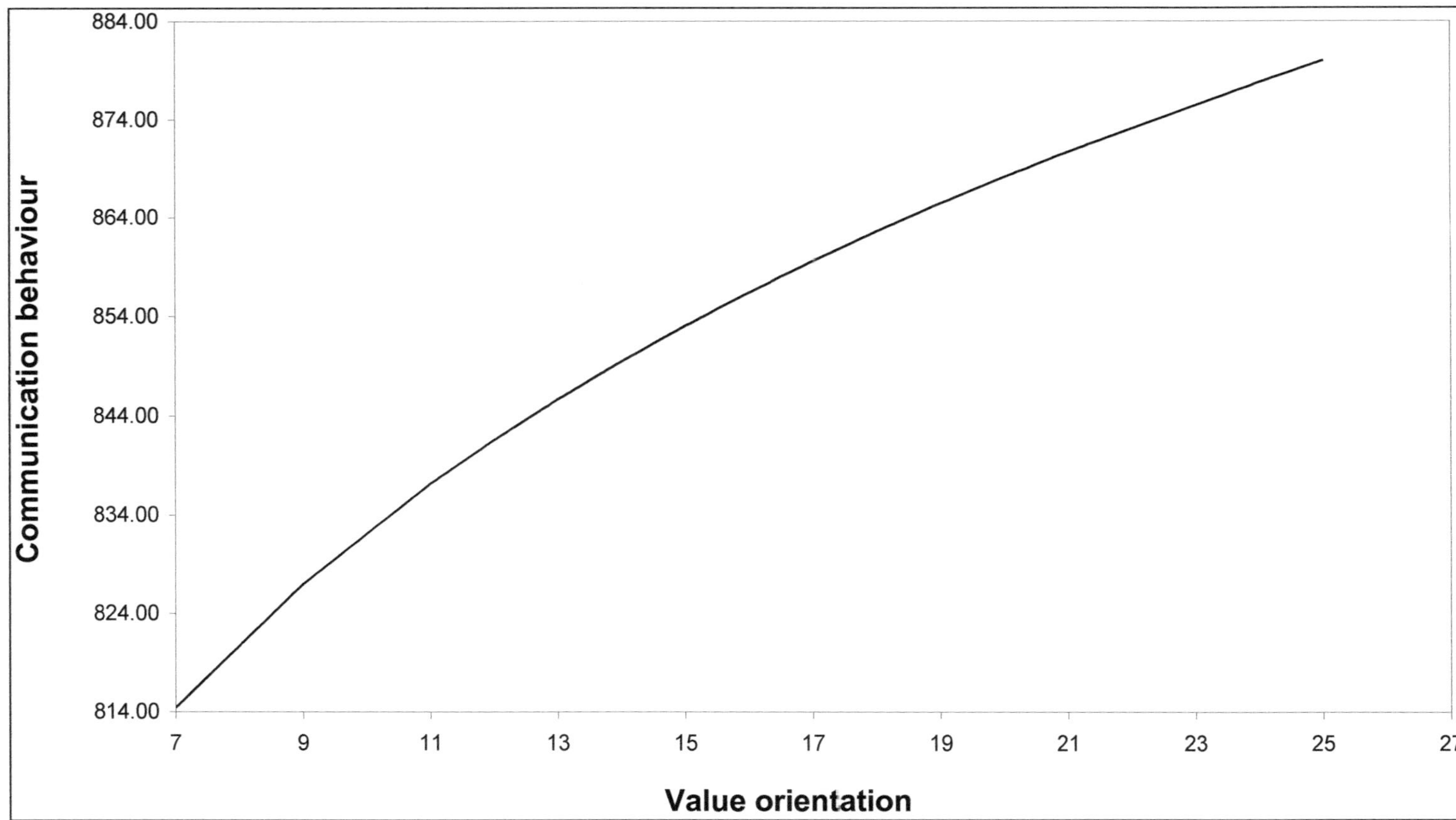

Figure 6: Exponential Model of Communication Behaviour (5).

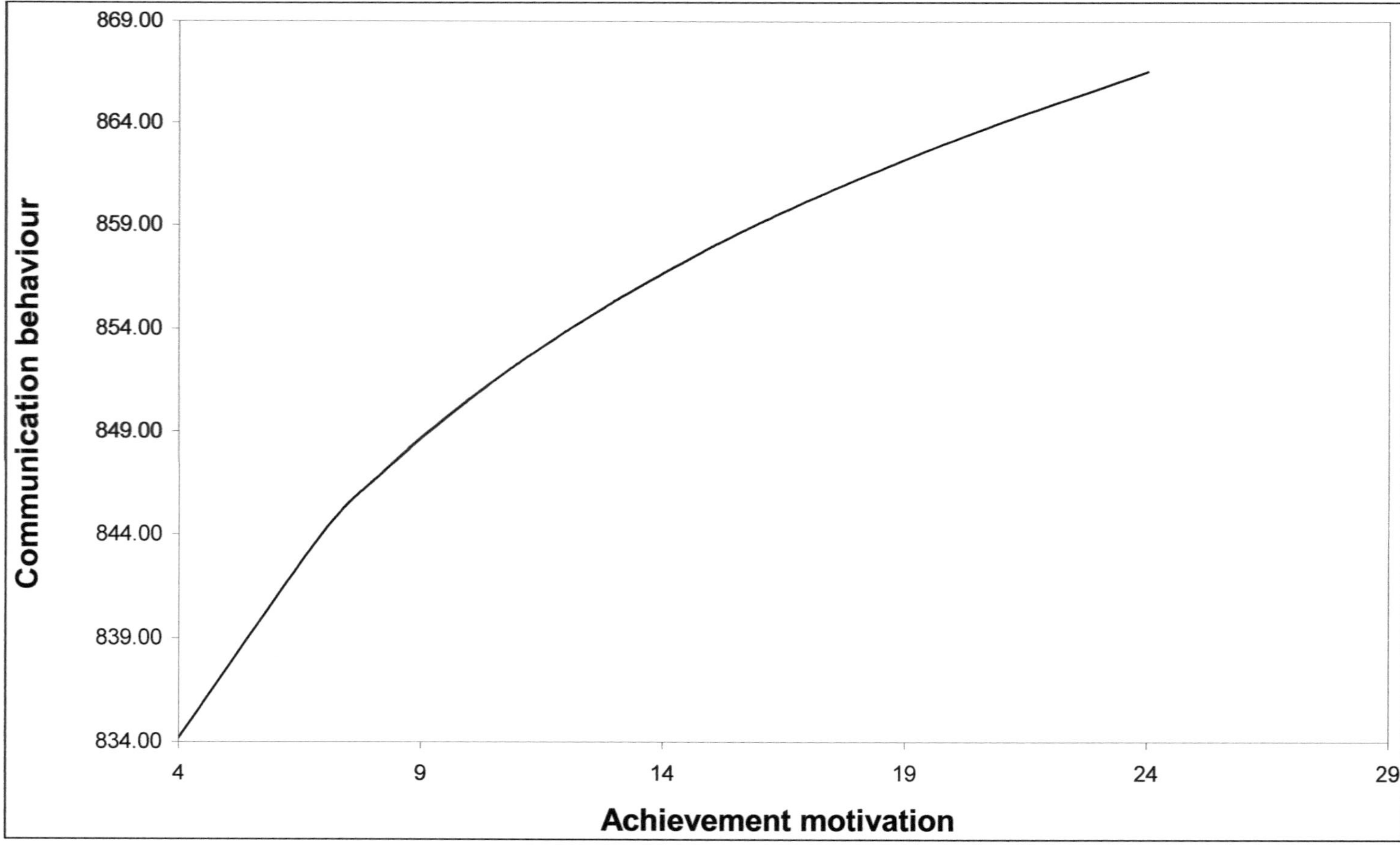

Figure 7: Exponential Model of Communication Behaviour (6).

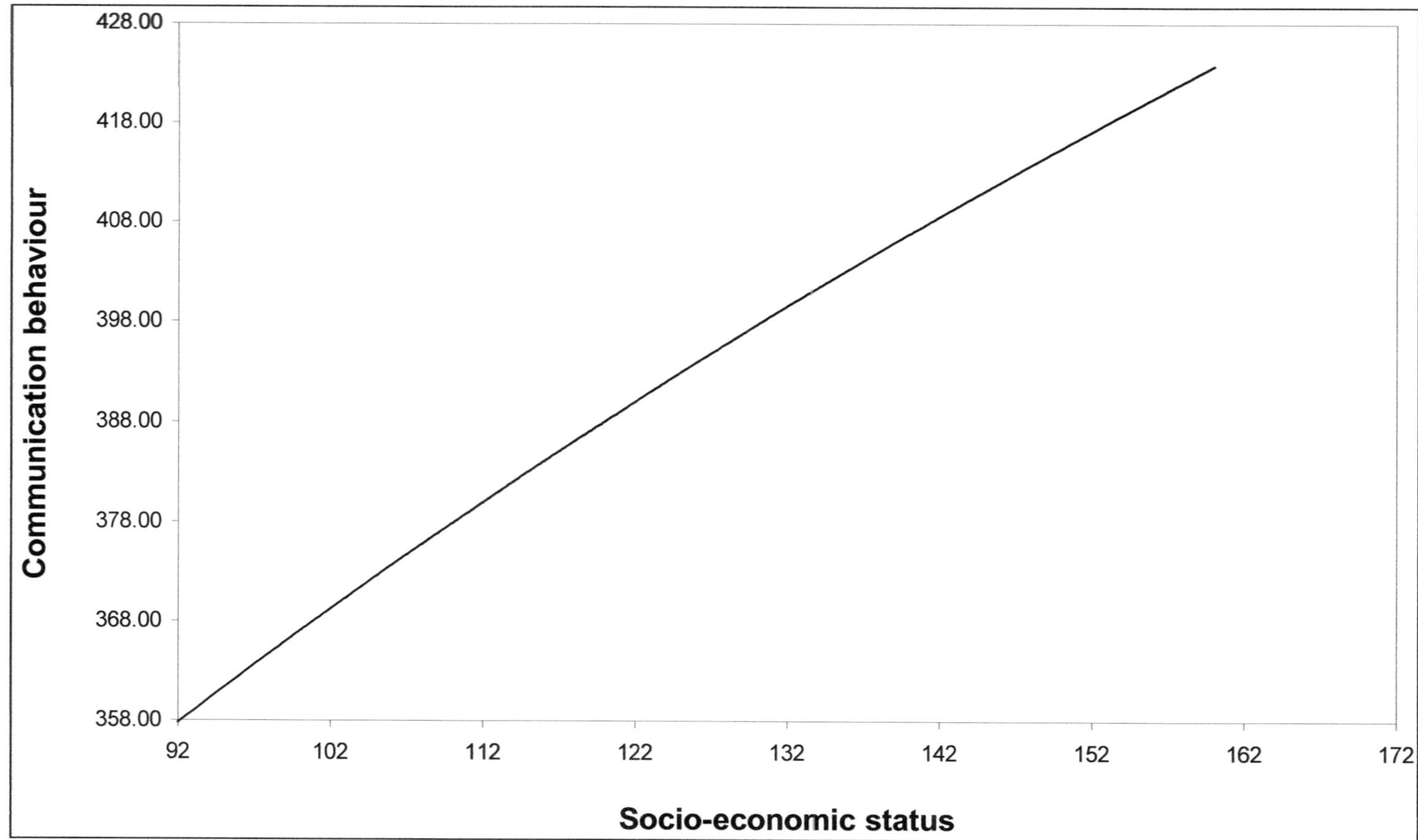

Figure 8: Exponential Model of Communication Behaviour (7).

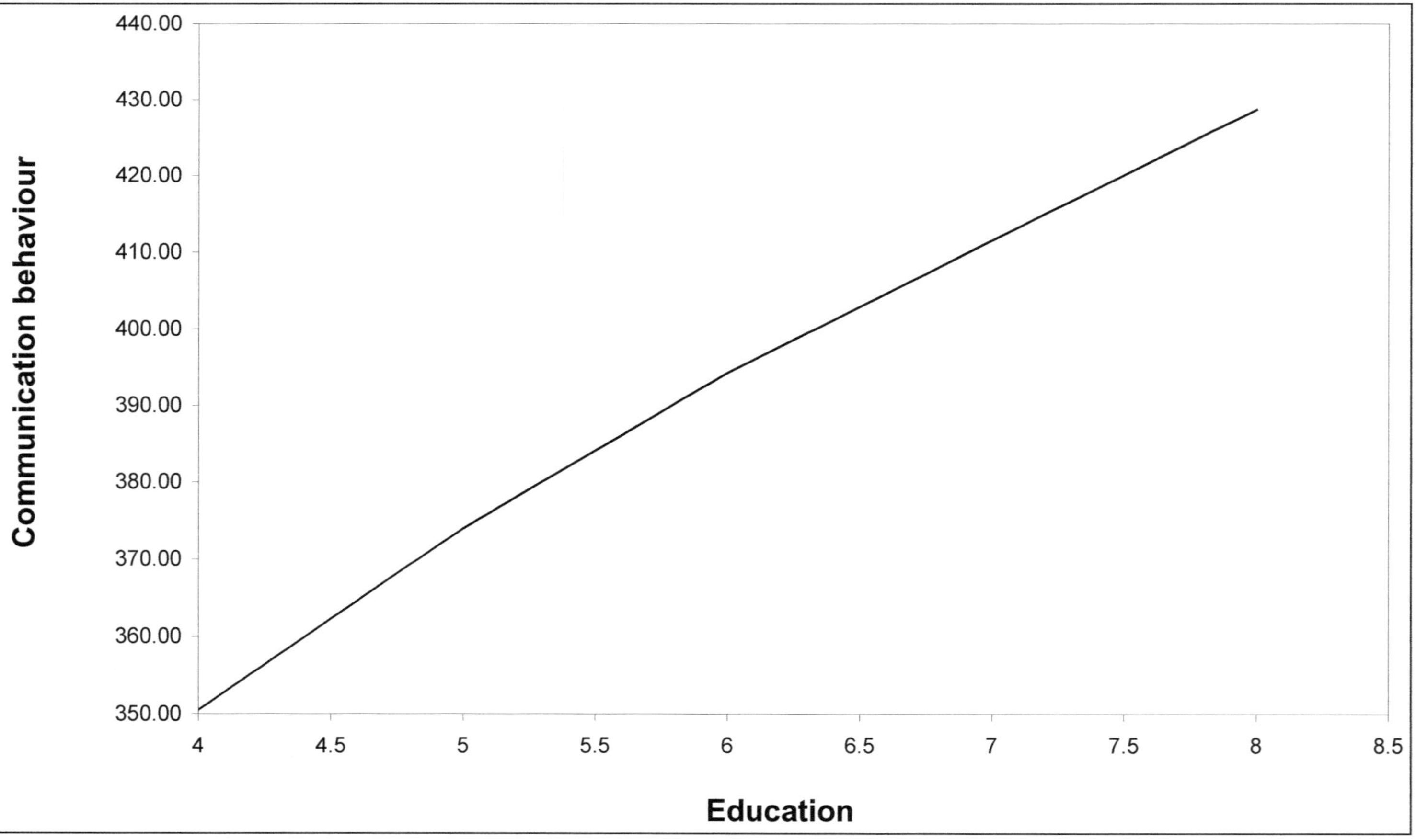

Figure 9: Exponential Model of Communication Behaviour (8).

orientation, lastly a saturation point reached after that the communication behaviour of the farmers did not increase.

From Figure 7, it can be expressed that with the increase in achievement motivation, the communication behaviour of the farmers towards farm education programmes of television increased with decreasing ratio. Ultimately a saturation point reached with the increase in achievement motivation, after that communication behaviour of the farmers towards farm education programmes of television did not increase.

From Figure 8, it is revealed that with the increase in socio-economic status, communication behavior of the farmers towards farm education programmes of television also increased with the decreasing ratio. Here, in this case the decreasing rate is so slow that it is not visible in this model. Ultimately a saturation point reached with the increase in socio-economic status, after that the communication behaviour of the farmers toward farm education programmes of television did not increase.

From Figure 9, it is depicted that with the increase in education, communication behaviour of the farmers towards farm education programmes of television also increased with the decreasing ratio. Here also the decreasing rate is so slow that it is not visible in this model. With the increase in education, communication behaviour of the farmers, ultimately reached to a saturation point after that the communication behaviour of the farmers did not increase.

It is concluded that value of R^2 found in case of linear function was lesser than that in the exponential function. So, exponential model (Cobb-Douglas Model) was considered as the best fit. In exponential model, with the increase in inputs (independent variables), output (communication behaviour) increases with the decreasing ratio. Ultimately a saturation point reaches after that the output communication behaviour of the farmers does not increase. Further, in this model all the regression coefficients were found to be significant whereas, in the linear regression model some of them were non-significant. It shows that all the independent variables were contributing significantly towards communication behaviour in case of Cobb-Douglas model.

Chapter 5

Summary and Conclusion

Communication is the core activity of human association in general and progress as well as development in particular. No human life can exist in isolation. A man can survive in society and the survival in society is possible with communication. Therefore, communication is identified as the oldest continued activity of human being since birth and goes on and on till death. More precisely, communication is the basic need of human being and web of society which makes the survival, growth, progress and development of man possible and holds the society intact and progressive.

Developmental information and technologies generated for the farmers are of no use unless reach to the ultimate users. It has been estimated that only about 30 per cent of the technologies are being received and used by the farmers (ICAR, 1988). It is further added that the information/technologies generated today reach to the entire ultimate users in about 20 years.

The use of present extension and communication technology system is based on the initiatives of the farmers - the receiver itself. This is possible only when the farmer is conversant with the knowledge of handling system, approach etc. about present communication technology system as well as the positive attitude towards the system. In view of the progressive farmers, its use is judicious as they have high level of positive communication behaviour. The positive communication behaviour has resulted the desired results in their agricultural profession.

It is beyond discussion and justifications that the communication behaviour is the basic activity of an individual through which the information is converted into its action for desired results. Katz and Lazarsfield (1955) considered communication behaviour as the respondents listening and reading habits. Rogers (1966) defined communication behaviour as the degree to which an individual is willing to seek information, use and advice. Newcomb *et al.* (1965) viewed communication behaviour manifested in sensitivity to information about the properties of the referent, the actualization of information such that the sender and receiver have more nearly equal information about the referent, understand the information, leading to change in attitude structure and in motivating satisfaction of goals.

Models are useful theoretical constructs that are frequently used in social science for explanatory purposes. They may be used to show the size, shape or relationship of various parts or components of an object or process. A model may also be useful in explicating the working of a system. According to Deutsch (1952) models perform an organizing function in explaining the relationship of one part to another and in giving us an idea of the whole system. At the same time, several authors, viz. Broadbeck (1959), Bross (1965) state that models have a somewhat ambiguous standing in the philosophy of science. Other view models as a brief concept from visualized ignorance to totally well-formulated scientific theories.

In rural development nothing is more important than the transfer of useful ideas from one person to another. The researches in agricultural sciences are of no use, unless they are communicated to the farmers in an effective manner in the shortest possible time. T.V. has emerged as a powerful medium of communication. Television is providing instructions and entertainment to far flung areas. While it provides sound, vision and movement, it can reach the largest number of people in the shortest possible time.

Therefore, keeping in mind the importance of T.V. in the field of agriculture, the present study entitled "Communication Model on Farm Education Programmes of Television in Kathua District of Jammu and Kashmir" was undertaken with the following specific objectives :

1. To study the communication behaviour of the farmers towards farm education programmes of television.
2. To find out the preference of television in comparison to other agricultural information sources.
3. To examine the association between selected independent variables namely socio-economic status, education, size of land holding, annual income, interpersonal interaction, innovation proneness, value orientation and achievement motivation with the communication behaviour of the farmers towards farm education programmes of television.
4. To identify the agricultural technology for farm education programmes of television using entertainment farm education format and ranking of programmes if available from television centres and respondents.
5. To develop an appropriate model for improving the communication behaviour of farmers.

The study was conducted in purposively selected Kathua district of Jammu and Kashmir. Out of 9 C.D. Blocks only 4 C.D. Blocks namely Ghagwal, Hiranagar, Kathua and Barnoti were selected randomly. Thereafter 20 per cent Gram Panchayats from each C.D. Block were selected randomly. Then a sample of 20 per cent villages was selected randomly from selected Gram Panchayats. Then a sample of 20 per cent farmers/respondents was selected randomly. In view of objectives of study, review of literature and discussions with academicians 8 independent variables socio-economic status, education, size of land holding, annual income, interpersonal interaction, innovation proneness, value orientation and achievement motivation were selected. The selected dependent variable was communication behaviour. Data were collected with the help of structured interview schedule specially developed based on standard scales with modification in light of objectives and selected variables of study. The collected data were tabulated, classified and statistically processed by using mean, percentage standard deviation, coefficient of correlation, linear regression and Cobb-Douglas function.

Findings

The important findings of the study are given below objective wise

1. To study the communication behaviour of the farmers towards farm education programmes of television.
   - (i) Majority of farmers were using television, radio, contacting extension personnel of C.D. Blocks, salesmen of agril. inputs, progressive farmers and local leaders as the main sources of information.
   - (ii) Majority of farmers used to evaluate the information by discussing with the progressive farmers, local leaders, neighbours and family members.
   - (iii) Majority of farmers stored the information by memorization and writing in general notebook.
   - (iv) Majority of farmers transformed the information by rearranging the important information as per their needs and rearranging the information in local dialect.
   - (v) Majority of farmers disseminated the information to their family members, friends and neighbours.
   - (vi) 16.00 per cent farmers had low communication behaviour towards farm education programmes of television.
   - (vii) 55.33 per cent farmers had medium communication behaviour towards farm education programmes of television.
   - (viii) 28.66 per cent farmers had high communication behaviour towards farm education programmes of television.
2. To find out the preference of television in comparison to other agricultural information sources.
   - (i) It was found that T.V. was ranked third in comparison to extension personnel of C.D. Block and salesmen of agril. inputs who were ranked first and second respectively. Radio was ranked at fourth place.
3. To examine the association between selected independent variables with the communication behaviour of the farmers towards farm education programmes of television.
   - (i) The communication behaviour of the farmers was positively and significantly related with the socio-economic status, education, size

of land holding, interpersonal interaction, innovation proneness, value orientation and achievement motivation. While annual income was positively and non-significantly related with the communication behaviour of the farmers towards farm education programmes of television.

(ii) Socio-economic status, education, size of land holding and interpersonal interaction were important variables for predicting the communication behaviour of the farmers towards farm education programmes of T.V.

4. To identify the agricultural technology for farm education programmes of television using entertainment farm education format and ranking of programmes if available from television centres and respondents.
   (i) Majority of farmers said that methods of using insecticides and pesticides and guidance about dairy should be shown in an entertainment-cum-education format.
   (ii) Majority of farmers liked to watch Krishi Darshan.
   (iii) Majority of farmers liked to watch methods of using insecticides and pesticides, guidance about dairy and guidance about poultry production in an entertainment-cum-education format.
5. To develop an appropriate model for improving the communication behaviour of farmers.
   (i) Cobb-Douglas model was considered as the best fit.
   (ii) Socio-economic status, education, size of land holding, annul income, interpersonal interaction, innovation proneness, value orientation and achievement motivation taken together were found positively significant and were important in predicting the communication behaviour of the farmers towards farm education programmes of T.V.

Implications of the Study

1. The communication behaviour of the farmers was greatly influenced by the farm education programmes of T.V. Thus, T.V. is an effective mass communication medium for dissemination of agricultural information. Therefore, more and more villages should be covered under T.V. network so that they could get the benefits of agricultural technology.
2. The findings of the study provide the information regarding the effectiveness of T.V. in comparison to other agril. information sources, to identify the agril. technology using entertainment education format and also the communication model which help to find out the important elements of affecting communication behaviour of the farmers. All this will help the planners to make the farm education programmes of T.V. more effective.

Suggestions for Future Research

1. The study was conducted only in four blocks of Kathua district of Jammu and Kashmir. The area of research could be extended further and sufficiently large number of samples should be studied to draw more valid conclusions.
2. A comprehensive study may be made to explore all possible factors that might influence the communication behaviour of the farmers.
3. Farm T.V. Programmes should be made in entertainment-cum-education form.

References

Ahmed, M.B. and Ali. M.M. (1998). Impact of farmers communication behaviour on types of information generated in BS farmer. *Annals of Bangladesh Agriculture*. Vol. **8**, No. 2 pp. 119-137.

Akhouri, M.M.P. (1973). Communication Behaviour of Extension Personnel : An analysis of Haryana Agricultural Extension System. Unpublished Ph.D. Thesis, Division of Agricultural Extension, IARI, New Delhi.

Ambastha and Singh, K.N. (1978). Intra-personal communication pattern of extension personnel, *I.J.E.E.*, **14**(1 and 4): 8-15.

Ambastha, C.K. (1974). Communication Pattern of Farm Information Development Extension and Client System in Bihar, A System approach, Unpublished Ph.D. Thesis, Division of Agril. Extension, IARI, New Delhi.

Ambastha, C.K. and Singh, K.N. (1978). Intra-personal communication pattern of Extension personnel, *I.J.E.E.*, **14** (3 and 4): 8-15.

Ani, A.O. and Kwaghe, P.V. (1997). "Sources of information on improved farm practices. a study of farmers in Unuahia zone of Abia state, Nigeria". *Agroresearch*, **3**(1-2): 1-10.

Arneja, C.S. and Singh, D.P. (1998). "A study of farm periodicals as source of farm information readers judgments". *Indian Jr. of Ext. Edu.* **34**(3-4): 143-146.

Aubel, J., Rabei, H., Mukhtar, M. (1991). Health workers attitudes can create communication barriers. *World Health Forum*, Vol. **12**, No. 4, pp. 46-471.

Babu, R. and Sinha, B.P. (1985). Communication behaviour of extension personnel with regard to modern rice technology, *I. J.E.E.*, **21**(3 and 4): 11.

Balak, Shiv (1999). Effectiveness of different sources of information for participation in rural agricultural development programme. Unpublished M.Sc. Thesis, C.S.A. Univ. of Agri. and Tech., Kanpur.

Basbao, Casole Anand Pease, Elizabeth M. (1992). Interpersonal communication motives and the life position of elders. *Communication Research*, Vol. **19**, No. 4, August, 1992, 516-531.

Bembridge, T.J. (1993). Farmer characteristics and contact with information sources in a Venda village implications for extension. *South African Journal of Agricultural Extension*, Vol. **22**, pp. 19-22.

Berlo, Qarid, K. (1960). The process of communication. New York: Holt, Rinehart and Winston Inc.

Bhangoo, S. and Kaur, A. (1994). Multimedia approach in adult learning. *Indian Journal of Ext. Edu.*, Vol. **XXX** (1 and 4): 86-90.

Bhople, R.S. (1958). Communication patterns in farm innovation extension acceptance and adoption by the farmers in Maharashtra State, Ph.D. Thesis, College of Agriculture. P.A.U., Ludhiana.

Blackie, E. (1986). Television and the education of adults. *Adult Education*, **59**(2): 147-152.

Boumon, M.P.A. (1999). The Turtle and the Peacock : Collaboration for pro-social change. *Wageningnen*. The Nethelands Wageningen Agricultural University, pp. 103-105.

Brown, W.J. and Singhal, A. (1949). "Ethical consideration of promoting Pro-social messages through the popular media". *Journal of Popular Films and television*, **2**(3) pp 92-99.

Browne, D. (1983). "Media Entertainment in Western World" In L.J. Martins and A.G. Chardhang (eds.). *Comparative Mass Media System*, pp. 187-208 NY, Longman.

Bruce, Rogers and Haskar (1971). "Education communication and agriculture change. A study of Japanese farm". Ph.D. Thesis (unpub.) comparative education centre, the University of Chicago.

Bruce, Rogers and Hasker (1971). Education, communication and agricultural change: A study of Japanese Farm, A Ph.D. Thesis, Comparative Education Centre, The university of Chicago.

Cassata B. Mary and Asante K. Molefi (1979). Mass Communication : Principles and Practices: Macmillan Publishing Co., Inc. New York Collier Macmillan Publishers, London. pp. 63-73.

Champawat, J.S. and Intodia, S.L. (1970). Sources of Information at the various stages of adoption of the weedicides, *I.J.E.E.*, **6**(1 and 2): 116-118.

Chandra, S. and Babel, K.S. (1997). A study of information available to the small, medium and large farmers about improved agricultural practices of Moong bean cultivation Annals. *Agri-Bio-Res.*, **2**(2): 77-80.

Chauhan, R.S. (1997). "Farmers response towards Farm Television Programmes in Jaipur district, Rajasthan". M.Sc. (Ag.) Thesis (Unpub.), S.K.N. College of Agriculture, Jobner.

China, R. (1986). Nigeria: Training and demonstration: A system to meet a need; *RRDC Bulletin*: University of Reading September.

Clark, R.C. and Akinbode, T.A. (1968). Factors associated with three Farm Practices in western state of Nigeria, *Research Bulletin-1*, Faculty of Agriculture, University of Lieife.

Dass, P.K. and Sharma, J.K. (1998). "Credibility pattern of different sources of farm information". *Journal of the Agricultural Science Society of North East India*, **11**(1): 121-123.

De Fleur, Mehin. (1966). The Theories of Mass Communication. New York: David Mckay Co. Inc.

Dhondyal, S.P. (1997). Farm Management: Economic Approach, pp. 213-216.

Dissanayake, W. (1984). Communication models and knowledge dissemination. *Media Asia*, Vol. **II**, No. 3, pp. 123-128.

Gaikwad, V.K., Tripathi, B.L. and Bhatnagar, G.S. (1972). Opinion leaders and communication in India villages: Centre of management in Agriculture; Indian Institute of management, Ahmedabad.

Gerbner, G. (1960). Towards a general model of communication. *Audio Visual communication review*, Vol. **14**, pp. 15-19.

Gupta, J. (1991). "Progressive use of communication media by marine fisherman". *Journal of Extension System*, **7**(1): 85-90.

Hai, Abdul, Srivastava, R.M. and Singh, R.P. Ratan (2003). Livestock farmers preference of communication medial and their use by extension workers in tribal, Bihar. *Indian Journal of Extension Education*, Vol. **XXXIX** (1 and 2): 31-34.

Hiranand and Jain, N.C. (1976). A study of extension methods used by block personnel for Extending Improved Agricultural Practices, *I.J.E.E.*, **3**(1 and 2).

Hovland, Carl L. (1964). Communication and Persuasion, New Haven, Yele Univ. Press.

Hutanawetr (1982). Socio-economic constraints in Rainfed Agricultural Production in Lower North East Thailand, Faculty of Agriculture, Khon Khaen University, page 25.

IARI (1980). *Research Bulletin* No. **35**.

Jain, Nemo, C. (1970). Communication patterns and effectiveness of professional performing linking roles in a Research Dissemination Organization: Ph.D. Thesis, Michigan State University, U.S.A.

Jha, P.K., Chauhan, J.P.S. (1999). Correlates of interpersonal communication behaviour of dairy farmers in north Bihar. *Journal of Dairying, Foods and Home Sciences*, Vol. **18**, No. 1, pp. 55-57.

Kadian, K.S. and Kumar, Ram (2003). Information processing pattern of dairy farmers of Kangra Valley. *Indian Journal of Extension Education*, Vol. **XXXVIII** (1 and 2) : 65-70.

Kaur, S. (1982). Comprehensive and use of information. A study on correspondence course for farm ladies in Punjab. unpublished Ph.D. thesis, Department of Extension Education, P.A.U., Ludhiana.

Khan, M.A. and Paracha, S.A. (1994). International 'communication network in diffusion of innovations at innovative and non-innovative villages. *Jr. of Rural Development and Administration*, **26**(2): 74-88.

Laharia, S.N. and Talukdar, R.K. (1986). Path Analysis of Factor influencing the Productivity of A.D.O., *I. J. of Social Work*, **XLVII** (3) : 225-265, October.

Lal, B. (2002). "Farm T.V. Programmes in Kathua district of Jammu and Kashmir". M.Sc. (Ag.) Thesis (Unpub.), S.K.N. College of Agriculture, Jobner.

Lasswell, H.D. (1948). The structure and functions of communication in society. In : *The communication of Ideas*. New York, Institute for Religious and Social Studies, 37 pp.

Leagans, J. Paul (1961). Characteristics of teaching and learning in extension education. In : Extension Education In Community Development. New Delhi : Ministry of Agriculture, pp. 171-194.

Martinez Ruiz, J., Sanchez Izquierdo, M.A. (1997). Communication as a means of making growers participants in sustainable development. Sustainable irrigation in areas of water scarcity and drought. *Proceedings of the International Workshop*, Oxford, U.K., pp. 14-22.

Mathur, P.N., Singh, K.N. and Lokhande, M.R. (1974). Sources utilization and rates of spread of information of HYV of wheat in farming community. *Indian Journal of Extension Education*, No. **10**, pp. 1-2.

Mc Quail, D. and Widnahe, S. (1981). Communication Models. New York : Longman Publishers Inc.

Murthy, T.K. (1965). "A study of predictive value of some factors of adoption of nitrogenous fertilizers and the influence of sources of information on adoption behaviour. Ph.D. thesis IARI, New Delhi.

Pandey, S.N. (1979). Communication Pattern under the T and V system of Agricultural Extension in Chambal Command Area Development Project of Rajasthan. Unpublished Ph.D. Thesis, Division of Extension, IARI, New Delhi.

Pandit, S. (1962). A study of Role of Age, Education and Size of Farm in Relation to Adoption of Improved Agricultural Practices, P.G. Department of Agril. Extension, Bhagalpur University, M.Sc. (Ag.) Thesis, pp. 14-152.

Parvanta, C.F., Gottert, P., Anthony, R. and Parlato, M. (1997). Nutrition promotion in Mali : highlights of a rural integrated nutrition communication programme (1989-1995). *Journal of Nutrition Education*, Vol. **29**, No.5, pp. 274-280.

Piotrow, P.T. (1994). "Entertainment Education: An idea whose time has come". *Population Today*, pp. 4-5.

Rajput, R. (1993). Sources of agricultural information : A study on Noopur block of Bijnor district in U.P. Unpublished M.Sc. Thesis. G.B.P.U.A. and T., Pantnagar.

Raju, V.T. and Rao, D.V.S. (1999). Economics of Farm Production and Management, pp. 121-124.

Rao, B.S.S. (1992). Television for Rural Development, New Delhi. India concept publishing company, 264.

Rastogi, A. (1995). Communication strategy: A need of sustainable development - JFM perspective. *Indian Forester*, Vol. **121**, No. 5, pp. 339-349.

Rathore, Rajendra (2000). A study on information needs and utilization pattern of the farm publications published by the Rajasthan Agriculture University, Ph.D. Thesis, G.B.P.U.A. and T., Pantnagar.

Reddy, Brya, H.N. and Singh, K.N. (1977). Analysis of communication pattern and procedures used by V.L.Ws, in India; in Cough and Chamela 9(ed.) (1984) Vol. **2**; (q.v.) Chapter 18, 221-235.

Reddy, S.K. and Kivilin, J.E. (1978). Adoption of high yielding varieties. Behavioural Science and Community Development, No. 2p. 121-122.

Reddy, U.V. and Sinuhal, H. (1987). Television in higher education, the Indian experience. *Media in Education and Development*, **20**(4): 128-137.

Rehman, M.Z.; Kashem, M.A., Mikuni, H. (1997). Farmers communication behaviour for technological intervention of Japan, Bangladesh cross-cultural study. *Jr. of Interacademicia*, Vol. **1**, No. 4, pp. 277-290.

Riley, J.W. and Riley, M.W. (1969). Mass communication and the social system. In Sociology Today. R.K. Merton *et al.* (eds.) New York : Basic Books, pp. 151-163.

Rogers, E.M. and Shefner-Rogers, C.L. (1996). "Evaluating the effect of an entertainment" with multi-media. Measurement Association Albuquerque, NM.

Sachchidanand (1972). Social Dimension of Agricultural Development, National Publishing House, Darya Ganj, New Delhi.

Sadaqath, S., Changiri, D.M. and Sunderaswami, B. (1998). Awareness and utility of agricultural publications among farmers. Maharashtra, *Journal of Extension Education*, Vol. **XVII**: 78-81.

Sandhu, A.S. (1973). "Relative efficiency of four methods of measuring credibility of farm information sources". *Indian Jr. Ext. Edu.*, Vol. **18**, 3-4, 74.

Sanoria, Y.C. and Singh, K.N. (1976). Communication patterns of Extension Personnel, System Analysis *I.J.E.E.*, **12** (1 and 2): 28-34.

Sarkar, A. (1997). "Impact of mass communication on rural people in relation to agricultural information". *Environment and Ecology*, **15**(4): 862-866.

Sauer, M. (1992). Nigeria and India, the use of film for development whispers in a crowd. *Africa Media Review*, **6**(1): 25-31.

Shannon, C. and Weaver, W. 1949. The Mathematical Theory of Communication Urbana. Univ. of Illinois Press.

Sharma, R.K. (1994). Farmers perception of constraints in milk marketing and measures for development of efficient extension system for milk marketing in rural areas. *I. Jr. of Dairy Science*, Vol. **47**, No. 8, pp. 674-679.

Sharma, V.P. (2003). Cyber Extension Approach for new Millennium. Media and Cyber Extension III. Reading Material : 1-21.

Sherief, A.K., Menon, A.G.G. and Baskaran, C. (1993). Kerala India : Communication behaviour of non-contact farmers. *Rural Extension Bulletin*, No. 3, pp. 33-35.

Singh, A.N. and Prasad, C. (1972). A study of characteristic expectations and listening behaviour of the farm radio-programme and its impact on acquisition of knowledge. Ph.D. Thesis Research in Comm., Division of Agri. Extension, I.A.R.I., New Delhi.

Singh, A.R. and Gupta, Pushpa (2000). Visual perception and comprehension of charts by tribal women. *Indian Journal of Ext. Edu.*, Vol. **XXXVI** (1 and 2): 21-26.

Singh, B. and Dubey, V.K. (1991). Use and Abuse of Pesticides by the Vegetable Growers. *Interaction*, **9** (1 and 2): pp. 113-138.

Singh, B.B. and K. Singh. (1997). Development Communication and Agricultural Extension : Interface. *Frontier of Ext. Edu.*, For 21st Century, *ISEE* : 84.

Singh, B.N. and Jha, P.N. (1965). Utilization of farm information in relation to Adoption of Improved Agricultural Practices, *I.J.E.E.* **1** (1): 34-42.

Singh, Bharat, Narwal, R.S. and Malik, J.S. (2003). Communication sources used by Extension personnel and farmers. *Indian Journal of Extension Education*, Vol. **XXXIX** (1 and 2): 26-30.

Singh, D. and Singh, M.R. (1971). Farmers perceptions of different credibility pattern of sources of information. *I.J.E.E.*, Vol. **VIII** (3 and 4) p. 78.

Singh, K.N. and Kumar, V.K. (1965). Result Demonstration Purpose, Process and Technique-1, *I.J.E.E.*, **1**(2): 92-101.

Singh, K.N., Rao, C.S.S. and Sahay, B.N. (1970). Research in Extension Education for Accelerating Development Process, *Indian Society of Extension Education*, IARI, New Delhi.

Singh, Mridula and Lal, Murari (1997). Population growth poverty syndrome and environment. Souvenir : 23.

Singh, N.P. and Singh, P.M. (1975). Rural telecast for development. An impressionistic model. *Economic and political weekly*, Vol. **X** (36), p. 14-34.

Singh, R. (1973). "Effectiveness, of communication media for rural auidence". An experimental study. Unpub. B.Sc. Thesis, Bombay.

Singh, R. and Sharma, S.S. (1973). A comaprative study of adoption of HYVP by farmers of T.V. and non-T.V. Villages. *I.J.E.E.*, Vol. **IX** (1 and 2). p. 92-01.

Singh, R., Tyagi, K.C. and Singh, R. (1992). A study of the communication behaviour of dairy farmers. Vol. **47**, No. 6, pp. 480-486.

Singh, R.S.P. (1997). Axiomatic approach to theory building in communication behaviour of farmers. Ph.D. Thesis, Instt. of Agril. Sciences, B.H.U., Varanasi.

Singh, S.N., Singh, K.N. and Pala (1976). Media utilization for various categories of farmers with varying socio-psychological characters, *I.J.E.E.*, Vol. **XII** (182), p. 38.

Singh, Y.P. and Pareek, V. (1966). Studies on Agricultural Communication in India, Research in Extension Education. *Indian Society of Extension Education*, New Delhi.

Singhal, A. and Rogers, E.M. (1999). Entertainment Education: A communication strategy social change. Lawrence Erlbaum Publisher London.

Sinha, P.R.R. and Prasad, R. (1966). Sources of Information as Related to Adoption Process of some Improved Farm Practices. *Indian Journal of Extension Education*, Vol.**2**, No. (1 and 2): page 86.

Sohi, J.S. (1978). A study of some selected factors affecting adoption of Dairy Innovation by Different Categories of Dairy Farmers in Milk Shed Area of Ludhiana Milk Plant, Unpublished M.Sc. Thesis, NDRI, Karnal.

Sparks, B.A. and Collan, V.J. (1997). Communication in the service provider - customer relationship : the role of gender and communication strategy. *Jr. of Hospitality and Leisure Marketing*, Vol. **4**, No. 4, pp. 3-23.

Srivastava, J.P., Rai, R. and Kumar, K. (1988). "Communication behaviour of field extension personnel under J and V System". *Indian Jr. of Ext. Edu.*, Vol. **34** (3-4): 133.

Subharwal, K. and Verma, S.K. (1997). Media utilization in National Literacy Mission Centers. *Communicator*, Vol. **XXXII** (1): 33-36.

Sunderswamy, B. (1971). Extent of Adoption of Recommended Practices and Information Sources Consulted by the Farmers in Respect of Hybrid Jowar Cultivation in Selected Taluka of Mysore District. Unpublished M.Sc. Thesis, Division of Agricultural Extension, IARI, New Delhi.

Talukdar, R.K. (1976). Farmers perceptions of the utility of farm broadcast from AIR, Gauhati. Unpublished M.Sc. Thesis, P.A.U., Ludhiana.

Talukdar, R.K. (1984). Productivity of Agricultural Development Officers in Haryana: A Factor Analysis Study, Unpublished Ph.D. Thesis, H.A.U., Hissar.

Tripathi, A. (1977). A study of technological gap in Adoption of New Rice Technology in Coastal Orissa and the constraints Responsible for the same. Unpublished Ph.D. Thesis, Division of Agricultural Extension, IARI, New Delhi.

Trivedi, G. (1963). Measurement of analysis of socio-economic status of rural families (A study conducted in community development block, Kanjhawala, Delhi State). Ph.D. Thesis, Divn. of Agril. Ext. IARI, New Delhi.

Veerasamy, S., Rao, D.U.M., Venkatesan, T. and Satpadhy, C. (1994). Distortion of farm messages at different levels. *Indian Journal of Extension Education*, Vol. **XXXIX** (1 and 2): 87-90.

Vora, Erika Wenzel. "The Development of concept Diffusion Models and Their Application to the Diffusion of the Social Concept of Race". Ph.D. dissertation, State University of New York of Buffalo, 1978.

Westley, B. and Machean, M. (1957). A conceptual model for communication research. *Journalism Quarterly*, Vol. **34**, pp. 31-38.

Wiens G. Elmer (2006). "Egwald Economics: Microeconomics Production Functions". http://www.egwald. com/economics/production functions.php.

Williams, S.K.T. (1969). Sources of Information on Improved Farming Practices on Some Selected Areas of Western Nigeria, *Bulletins of Rural Economics and Sociology*.

Williams, S.K.T. (1969). Sources of Information on Improved Farming. Practices on some selected areas of western Nigeria, *Bulletins of Rural Economics and Sociology*, **4**(1): 8-30.

Index

www.ingramcontent.com/pod-product-compliance
Ingram Content Group UK Ltd.
Pitfield, Milton Keynes, MK11 3LW, UK
UKHW021957270726
14060UKWH00002B/551

9 789386 07122